DEBATES OF THE HMOLPEDIANS

A book by David Bossens

DEBATES OF THE HMOLPEDIANS

Copyright © 2013 by David Bossens
All rights reserved. This book or any portion thereof
may not be reproduced or used in any manner whatsoever
without the express written permission of the publisher
except for the use of brief quotations in a book review.

Published by Lulu, inc.

ISBN 978-1-291-24164-8

[URL] *http://www.lulu.com/spotlight/bosszus*

Table of contents

1. What is the Hmolpedia? 3-4
2. Some Challenges and Suggestions 5-11
3. The long life debate 12-62
4. Intelligence, consciousness and free will 63-92
5. Other interesting people at the Hmolpedia 93-96
6. Morality 97-107
7. Final thoughts 108-111

1. What is the Hmolpedia?

The website

Today, a field called 'human thermodynamics' is brought into the spotlight, much of it owed to Libb Thims, an American electrochemical engineer, founder of the Hmolpedia. The Hmolpedia is a niche website dedicated to the relatively unknown subject of human thermodynamics, or more generally, applying physics and chemistry to explain human existence. This wiki counts over 2600 articles, and allows people all over the world to discuss these ideas. People can also register for free and work on editing the wiki. The website can be found at eoht.info.

Notable concepts and theories

Elective Affinities: A 1809 novella written by Johann Goethe, which for the first time in human history, symbolically portrayed social relationships, $A + B \rightarrow AB$, and related these to the elective affinities by Bergman : $A = T\Delta S - \Delta H$. Drawing on these, human thermodynamics relates Gibbs free energy (which equals –A) to human relationships.

Animate perspective: Libb Thims holds that life is a defunct scientific theory. Oftentimes, the Hmolpedia will refer to Sherrington who said 'Chemistry does not know the word life' or Tesla who said 'There is no thing endowed with life'. In this view, things that were previously considered 'alive', are better identified as having higher levels of atomic reactivity, animation, and prolonged and driven bound state existence. The prolonged bound state causes different particles to adhere so that these behave as one whole. Only via the use of energy can these particles be splitted.

Physics-based morality: Thims, drawing on the work of Goethe, adheres to a morality based on physical principles, rather than the list of prescriptions the average person adheres to – don't steal and don't murder for instance.

The human molecule: Thims, with Jean Sales as a predecessor, considers the human as one molecule. Evidence he reports is the Ecological Stoichiometry study, which calculated a gross elemental formula for the human body. Similarly, Thims has calculated such a formula.

Human molecular orbitals: The human molecular orbitals theory acknowledges that the wave-particle duality of matter, as conceived by DeBroglie, might apply to larger molecules – such as the 'human molecule'. A molecular orbital is the solution of the Schrödinger equation that describes the ninety percent probable location of an electron relative to the nuclei in a molecule and so indicates the nature of any bond in which the electron is involved. Electrons (and possibly also molecules) act as both waves and particles, tending to have orbital motions in their trajectories.

Hmolpedian: Person who is a frequent visitor of the hmolpedia. Has interest in hmolpedian subjects such as life, free will, intelligence, and for all the application of physical science to human behavior.

Interesting people at the Hmolpedia

Libb Thims:

American electro-chemical engineer ; founder and main editor of the Hmolpedia ; founder of the Institute of Human Thermodynamics; author of *Human Chemistry* and other books on the subject of human thermodynamics. Is a strong atheist, adheres to a physics-based morality and considers himself a Goethean revolutionist.

Georgi Gladyshev: professor of physical chemistry and former chief of the laboratory at the Institute of Chemical Physics Academy of Sciences in Moscow, a position that he held till 2005. Gladyshev has published a number of papers on the physical chemistry of biological evolution and on the evolution of planetary systems. He is also noted for coming up with a thermodynamic method to identify foods that are healthy.

Jeff Tuhtan: holds a PhD in Engineering and is employed as an ecological engineer; is moderator and frequent editor at the hmolpedia.

DMR Sekhar: Chemical engineer with specialization in mineral process engineering. Did PhD on processing of sulfide minerals. Working in mineral processing plants since 1979. Noted for his Genopsych theory, and for holding awards related to mineral science and mine engineering.

And of course, **David Bossens**, completing MSc Psychology, theory and research, founder of BossensNonFiction.com, and author of this amazing book...

The five featured hmolpedians...

From left to right: Libb Thims, Georgi Gladyshev, Jeffrey Tuhtan, DMR Sekhar, and of course David Bossens.

2. Some challenges and suggestions

Predictions and applications? Gottmann as a case study.

While at the surface one would expect an approach to human behavior via equations and chemistry is more precise in its predictions, for now, it seems like this approach is poor on predictions. Proponents may say to take a look at the Gottmann study.

At first glance, it seems some successful predictions have been made with regards to marriage. Hmolpedia gives:

Through low-motion video recordings of human couple interactions, Gottman was able to develop a formulaic methodology that is able to "predict", with 94 percent accuracy, which couples will divorce in the long run and which will stay bonded in the long run. Gottman's models, however, only loosely employ thermodynamics logic. In 1999, for example, he gave his view that:

"Something like a second law of thermodynamics seems to function in marriage—that is, when marital distress exists, things usually deteriorate (entropy increases)."

He established ratios of positive to negative moments, and established that ratios of 5-to-1 are a stable bond, whereas ratios of 5-2 or worse, cause instability, which results in 'debonding'. As said, however, this loosely employs thermodynamic logic, and in fact, can be done without any reference to thermodynamics. Also, note that 'positive' or 'negative' are somewhat vague criteria. Many moments have multiple feelings associated with them. Another criticism, somewhat embarrassing to Gottmann no doubt, is that he didn't predict anything at all: he had the results and consequently made a model. No prediction.

This kind of research is far of, and will – in my opinion- always be far off in predicting the full behavior and choices of like-intelligent species: you can only predict that what is not ahead of you in terms of intellect, unless of course we make an extremely intelligent AI do the predicting for us. Still, those predictions are easier to make via macro-structural laws than via micro-structural laws.

Why does the human thermodynamics approach give no predictions or applications? Some proponents may attribute this to a delay caused by a violent repression by religious and anthropocentric thinkers. Partly, I must agree: there has been much reaction to notions such as determinism and the fact that we are not special, that we do not have a 'soul', and other such matters. Partly, I must disagree. There are other paradigms that should get the same response, since biological laws governing human reactions such as love, violence, etc are studied in the field of biological psychology. Yet these fields get much funding, and these fields have achieved much predictions – I am thinking about : stimulating specific neural areas will result in a specific behavior ; damaging a specific part of the brain will disable your behavior in specific ways ; inserting a brain chip against Parkinson's will cure one instance of such a disability.

However, I must moderate my criticisms: we mustn't expect from a new science what is not achieved by an old science immediately; there are lots of areas in which psychology and neurology only hypothesizes further and further without having any real life applications. An example of a prediction that *can* be made is for instance that a human molecule – I explain in the last chapter in what way this can be a correct term – will be able to be localized with a certain probability in a certain place. This approach is the human molecular orbitals theory. However, we don't actually need the human molecular orbitals to realize this. Rather, it seems the realization of the human molecular orbitals came after the knowledge that we can be localized in a certain place with a certain probability. This aside, we must indeed recognize that most of human science does not predict anything – but rather describes. This is most likely because humans are intelligent – which means that there are a large number of variables involved in behavior.

Overly complex? What about historical approaches?

I once said that human thermodynamics "is an interesting philosophy, a change of paradigm, perhaps even having a broader explanation for human behavior than Darwinian evolution." I speculated this way, because, if we were to find a higher level law (up the scale of hard science), we could explain why biological laws exist - since biology is but applied chemistry, and chemistry is but applied thermodynamics. However, I must somewhat revise my opinion, since Darwinian evolution is largely described in historical terms; To fully describe everything via thermodynamics would make things somewhat undesirably complex, if not render useful research impossible. For instance, if we say that 'bipedalism allowed for the broader gaze on the environment, technological manipulations by hand and better running,' then we don't need any chemical equations – it is in fact both undesirable and impossible to do so.

What about cognition and memories, and the brain?

Of one thing I am sure: we cannot fully predict systems that are equally or more intelligent than we are. Reason: I have lined out in other chapters how you need to be vastly more intelligent than the beings you like to predict. But in this case, once we know the rules, we might adapt to them. But even without all these, it is still easy to see that human thermodynamics disregards any cognitive factors. Although I am sure that everything in a sense is chemical, people are determined by their previous relations and the content, potential and inclinations of their own neural networks and capabilities. Behavior is too complex to predict via just a few equations.

We can of course make the argument that with a few simple rules, there can arise very complex behaviors over time and this is true. But, to make any prediction, in this case, you would have to be able to replicate every given memory and capacity of each individual to predict the behavior of this human.

Using cellular automata

Behavior is complex and I think using simple equations will not yield much fruits in terms of predictions. There are too much sensitivities to initial conditions, too many (unknown) factors and lots of undiscovered, or unintegrated laws. Furthermore, behavior although of course subject to basic physical laws, can be most directly and easily be predicted by the activities of the brain. But how do we predict the brain?

One solution which may be useful for human thermodynamicists, but of course for any scientist, is the use of cellular automata. A cellular automaton consists of a regular grid of *cells*, each in one of a finite number of *states*, such as *on* and *off*. For each cell, a set of cells called its *neighborhood* (usually including the cell itself) is defined relative to the specified cell. An initial state (time $t=0$) is selected by assigning a state for each cell. A new *generation* is created (advancing t by 1), according to some fixed *rule* (generally, a mathematical function) that determines the new state of each cell in terms of the current state of the cell and the states of the cells in its neighborhood. When John Conway developed his 'Game of Life', scientists began to realize that reality resembles such a cellular automaton.

It seems feasible to predict the behavior of complex systems if we insert the thermodynamical and other physical equations in such automata. The idea here proposed is to insert simple fundamental rules (for instance: behavior A is fit) into cellular automata, in much the same way as Hiroki Sayama, by using Langton's self reproducing loop, giving the following result:

"We succeeded in transforming the SDSR loop into an actually evolving one in a simple deterministic 9-state 5-neighbor CA space, by enhancing robustness of its state-transition rules, besides a slight modification of initial configuration of the loop. The experiment with the improved loop, named *evoloop*, met with the intriguing result that the process of spontaneous evolution emerged in the CA space, where loops varied by direct interaction of their phenotypes, fitter individuals were naturally selected, and the whole population gradually evolved toward the fittest species."

Applying the definition of life that is agreed upon

Libb Thims proposed online that the idea of Darwin's 'dark pond' which suddenly sprang to 'life', is not a very coherent one. Indeed, it would be much more credible and parsimonious to have as a rule that molecules bond and form ever more complex beings. 'Life' as Libb sees it, is meant as a physical force present in biology that is not present in non-biological things. This of course everyone then has to disagree with: 'life', as in a soul, a life force, *did not* suddenly sprang into existence, neither will it *ever* have existed – perhaps some religious people would disagree. No biologist defines life this way. What does constitute life is discussed later.

Try to synthesize the human molecular formula

Some criticism I have with regard to the human molecular approach by Thims is the following: suppose we recreate a human molecule
$C_{E27}H_{E27}O_{E27}N_{E26}P_{E25}S_{E24}Ca_{E25}K_{E24}Cl_{E24}Na_{E24}Mg_{E24}Fe_{E23}F_{E23}$ $Zn_{E22}Si_{E22}Cu_{E21}Be_{E21}I_{E20}Sn_{E20}Mn_{E20}Se_{E20}Cr_{E20}Ni_{E20}Mo_{E19}Co_{E19}V_{E18}$. If we would attempt to synthesize this formula, would it suddenly be a full-grown human?

I think not. Perhaps this may prove to be one falsifiable prediction of the 'human molecule hypothesis'? If we synthesize this formula, and we have an outcome different than a human being, then it should be clear that we are not 'just' a molecule, but that other conditions – besides molecular composition – should be considered. Or better, perhaps the assumption that many molecules form a giant molecule with precisely the same rules is untrue. If the prediction would be falsified, that does not necessarily mean that molecular composition is not key: more likely, I think it to be the case that it is not accurate to look at the human as *one* molecule, rather, I would think the human is a conglomerate of many molecules which are not necessarily bonded covalently. In the last chapter though, I will discuss why the human molecule may be a correct view, under a slight – but perhaps very useful - alteration of the definition of molecule.

Otherwise, if we had the chemical formula, then we should arrive at a human being by just synthesizing the formula in the lab – I have a suspicion that this will not be the case. It somewhat reminds me of a film where it was depicted that 'here is a human being', when clumping together water, blood, calcium, etc and then packaging it with a skin coat.

So in essence, it seems to me, that we are not *one* molecule, but rather, that we have *many molecules which are to be considered separately*. For instance, stomach acid is a very important tool in digestion, which also should be considered separately since we do not suddenly die, or even, become something 'not human' if we do not have it. Of course, we will be in trouble if its production fails, but still, the main ingredient of a human is not the fixedness of molecular composition, but the fixedness of genetic composition. So, try to synthesize a human with the formula above, and let us know whether the human molecule is a sufficient principle for making all what a human is.

My guess is that it is not. I consider DNA to be the most important characteristic to create a human being – and I'm sure many would agree, since after all, we grow larger and larger from just a single cell that contains half the DNA of the father and half the DNA of the mother. This single cell contains DNA that characterizes what we will become – a few errors in the genetic code have been show to lead to severe deviations. Genes can predict whether it will be a boy, or whether it will be deformed, or etc. Genes are indeed important to synthesize a human, as evidenced with cloning. My bet is that we cannot synthesize a human by just synthesizing the human molecule – a zygote with viable genes, however, has shown to give definite results in creating humans. I'm not sure what this means for the status of the human molecule, but I'm sure that the human molecule approach is *one out of many approaches*, each with their own validity and range of applicability.

Also - perhaps more important a factor for not arriving at a human by using some rough formula - the chemical formula cannot logically be the same for every person, since we all have somewhat different inclinations, appearances, and heights and widths, weight and bones, and genes. This cannot logically follow if we are subject to the same laws, and are completely the same in terms of molecular structure. But Libb of course acknowledges this as well that it is just the mean human molecular composition. So synthesizing a human molecular formula to produce a human would not prove to be counter-evidence for the human molecular formula – but rather underscore the importance of DNA.

Seeing the human as a whole is not always useful

Problem with the 'cell-as-a-molecule' and 'human-as-a-molecule' approach is that basically: here is the molecule, what do we do know? If we would look at the cell as composed of separate parts with specific functions (organelles) such as Golgi complex, mitochondria, phospholipid bilayer, DNA, then we would quickly see that the approach of a cell-as-a-molecule seems somewhat functionless, since we have no information about the telomeres and the sequences on DNA, energy production in mitochondria, etc. Why shouldn't we look at the function of these separate organelles? For instance, the production of ATP is an interesting property to know of by relating it to energy usage.

So, often it serves useful to – if molecule is a correct term for human – analyze the human in terms of its atoms, and to consider those atoms as molecules comprising even smaller atoms.

Molecular orbitals theory

The extrapolation of molecular orbital theory to the study of the structure, formation, and dissolution of chemical bonds between human molecules is called human molecular orbital theory. When human movement, over the surface of the earth, is viewed at a time-accelerated pace, such as viewing the total weekly, monthly, or yearly movements of one person, via for example GPS tracking, in a sped-up five minute video clip, one begins to see an orbital picture of human movement.

A molecular orbital, by definition, is the solution of the Schrödinger equation that describes the ninety percent probable location of an electron relative to the nuclei in a molecule and so indicates the nature of any bond in which the electron is involved. In simple terms, it is understood that electrons (and molecules) act as both waves and particles, tending to have orbital motions in their trajectories.

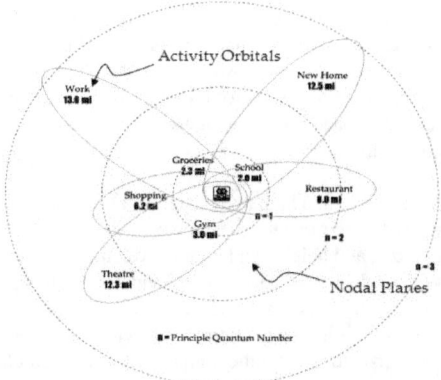

One interesting view, in this line of reasoning, I found the "molecular orbital hypothesis", where human molecules enter into each other's orbit to get attracted to each other, gradually getting more orbital overlap (stable relationship), and eventually a fusion takes place (moving in together). However, how is this hypothesis going to make a falsifiable hypothesis?

One other criticism in the assumption that macromolecules attract is the following. Suppose we make a human without any sexual/reproductive functions, no hormones etc. Shall this person be attracted to a human? He may 'like' some persons. But he sure will never be 'sexually attracted' to another partner. And this was what 'Elective Affinities', as conceived by Goethe, should mean: two people, usually of the opposite sex, get attracted to each other by human molecular attraction. However, the molecular orbital only describes location – this says nothing about the nature of the interaction. It only specifies that the person will be literally – that is, spatially – attracted to the person, community, place, and so on.

Integration with neuroscience and AI

Other methods have proven much worth. Especially the neurological approach to human sciences has been successful : stimulating one area and clearly relating it to some particular behavior implies much predictive power in the approach. Another such method is AI. Coincidentally, the creation of artificial intelligence seems to be in line with the reverse engineering of the human brain and human behavior, and thus the very creation of non-human intelligence implies more insight into our brain. Thus, although possibly being a truthful philosophy, with simple assumptions, there is more work to do in terms of predictions. This kind of research has a long way to go.

Memory, intelligence, movement and sensory capacities are important in behavior. A framework that consoles these areas of research with the investigations of the human molecule and the human molecular orbitals seems recommended.

The variability of bonds – on marriage

Does one marry the first woman one falls in love with? On the Hmolpedia, Libb describes marriage as establishing a human molecular bond. However, what about arranged marriages? What about when distance increases – goes to work for example? Is the bond re-established daily? Then marriage is re-initiated every day? What about all those other human molecular bonds? Is it not more plausible to assume rational choices in mating? For instance, SES, shared interests and complementary tasks? What about visual attraction ? Quite clearly no human molecular bonds are needed for arousal, since people get aroused pretty quickly by computer screens presenting naked people of the opposite sex.

Such questions are easily erased when you realize that the human is a molecule – indeed, in our expanded definition of molecule, which allows non-covalent bonds as well – whose 'atoms' are connected by photon-electron-bindings, and when you realize that the two atoms of a married couple are connected by photon-electron bindings, most likely non-covalent bindings. This way, bonds can be broken and re-established quickly.

All evidence is pointing to the fact that the wave-particle duality indeed applies to larger molecules. I refer to the work of Thomas Juffmann, who demonstrated wavelike behavior for a structure of more than 100 atoms.

Some reasons why it seems promising

Numerous aspects of this approach are promising: 1. Potentially very definite predictions, in terms of probability or other quantitive measures. 2. Another method to supplement existing scientific methods 3. A philosophical framework that acknowledges determinism, and thus jettison notions as free will, the soul or a life force – notions not supported by hard science.

Quick terminology

Before starting the debate, let us introduce some other important terms.

The quantity called *"free energy"* (dG) is a more advanced and accurate replacement for the outdated term *affinity (A)*, which was used by chemists in previous years to describe the *force* that caused chemical reactions. Do note that its direction is opposite: dG= - A.

Wikipedia defines Gibbs energy (also referred to as ΔG) as the chemical potential that is minimized when a system reaches equilibrium at constant pressure and temperature. The Gibbs free energy, originally called *available energy*, was developed in the 1870s by the American mathematician Josiah Willard Gibbs. In 1873, Gibbs described this "available energy" as

the greatest amount of mechanical work which can be obtained from a given quantity of a certain substance in a given initial state, without increasing its total volume or allowing heat to pass to or from external bodies, except such as at the close of the processes are left in their initial condition.

In the hmolpedia we read that the detailed understanding of the "state of equilibrium" and the various "criterions" for this condition, as system influences change, such as with particle (chemical species) movements in and out of the system, gravitational effects, osmotic effects, electromotive force effects, etc., were laid out in the 1876 publication *On the Equilibrium of Heterogeneous Substances* by American engineer Willard Gibbs. The term "state of equilibrium" was first mentioned in the opening paragraph to his abstract, where he notes: "It is an inference naturally suggested by the general increase of entropy which accompanies the changes occurring in any isolated material system that when the entropy of the system has reached a maximum, the system will be in a **state of equilibrium**."

When the *entropy* is maximal, it means there is no thermal energy left to do useful work. Hence some have proposed that death equals maximum entropy. The debates starts off with this matter of death and continues about its antonym: life.

3. The long life debate

Nothing is at equilibrium; everything is animate.

David: *Can we rightfully say that living beings are at an thermodynamic equilibrium - no free energy, and that at death we have reached the maximum entropy, causing us to go [into] equilibrium?*

Libb: *Firstly, there is no such thing as a "living being", that is a religious-mythological term. As to "maximum entropy" this is a very used and abused term. Mathematically, is simply means that the equivalence value of all uncompensated transformations", symbol, N, reaches its maximum positive value and the system stops transforming or changing state. All the order, disorder, life, death, etc., ideas are latter speculative appendages to the original version of what entropy increase means: namely that caloric is not a conserved property, some of it gets irreversibly transformed into work (movement of bodies by a force though a distance).*

David: *About the equilibrium: our 'human molecule' remains relatively stable, is that then not a sign of near equilibrium? or is it the opposite: can we say that, for example, my table seems to have no chemical interactions at all, and being in almost perfect equilibrium, while the human molecule is constantly changing, reacting, feeding?*

Jeff: *Nothing in the contemporary understanding of the known physical universe is at equilibrium. Thermodynamic equilibrium requires simultaneous mechanical, chemical, and thermal equilibrium. A rule of thumb regarding equilibrium is "if your system appears to be at equilibrium, then you are looking at the wrong scale."*
An example, if your glass of water appears to be at equilibrium at the scale you are observing (measuring) the rate change of its properties (mass, heat transfer, etc.) then you can be sure that if you look at the system at a finer scale over a shorter time interval (e.g. milliseconds), or at a longer time scale (e.g. years), that change will occur and thus it will be seen that the system was, is, and never will be truly in thermodynamic equilibrium.

Regarding a cellular system, only inanimate ("dead") systems can be considered to approach macroscopic (stuff we can see with our own eyes) equilibrium. Animate ("living") systems require external forcing (energy input from their surroundings) to maintain their motion. Humans are a very, very complex type of animate system, a result of billions of years of chemical evolution. Just because we may not be "alive" in the conventional sense does not mean that we have much to learn and appreciate about what and how we are.

David: *"Regarding a cellular system, only inanimate ("dead") systems can be considered to approach macroscopic (stuff we can see with our own eyes) equilibrium. Animate ("living") systems require external forcing (energy input from their surroundings) to maintain their motion."*
But this macroscopic equilibrium of inanimate matter is somewhat, but not entirely, indicating equilibrium at chemical, mechanical and thermal level then? whereas we know animate matter (living things) are not even in macroscopic way remotely in equilibrium (although the molecular structure remains somewhat stable) and thus not in chemical, mechanical or thermal equilibrium.

Jeff: *Inanimate systems are those which are not subject to external forcing. Your argument is correct in that actually all real systems are subject to some interaction with their surroundings, and thus if we view the full spectrum of scales of the really small and fast, the really huge and slow and everything in* between, then everything, all the time, can be viewed as a type of animate system!

The distinction between the "animate" and "inanimate" really is a sort of scientific parsing of the terms "living" and "nonliving" which deal with macroscopic systems. Other than to try and explain to people (mostly other scientists) that the commonplace notion of life (CNL) can be explained in modern science terms, there is no difference between the two.

So since these "big" macroscopic things are induced to move (phototaxis, gravitaxis, etc.) then these animate systems are not at equlibrium. Physics even allows a system to be in mechanical equilibrium if its motion is fixed at a constant rate (velocity with respect to some inertial reference frame), but if it is jiggling, jumping, or darting around, then it is not at thermodynamic equilibrium because the requirement of simultaneous mechanical, chemical, and thermal equilibrium is not satisfied.

David: ok, I will summarize what I now think to be the case:
-thermodynamic equilibrium is satisfied only when mech-, chem-, and therm- equilibrium are satisfied
-animate structures differ from inanimate in the sense that they differ on the macroscopic level
but it is still not clear to me if it was correct for me to say if:
-animate structures usually differ from inanimate structures by being LESS in equilibrium, since animate structures have much fluctuations in mechanical, chemical, and thermal sense, whereas for example a table will always have mechanical equilibrium, and chemical as well(?) (but not necessarily thermal)

Jeff: Your first point is correct. The second two points really only matter depending on with whom you are conversing. All systems are animate because in some way (tiny and fast, or huge and slow, or anything in between) they are interacting with their surroundings. So to be technically correct you could just say "animate = life" and everything in the universe is "alive."

The importance of this discussion is big, since via this discussion we realized that everything is 'animate' in the sense that nothing is in equilibrium. This has nothing to do with any 'panbioism', but rather it refers to the fact that there is no thing in equilibrium, no thing that is not moved by external forces. When either chemical, mechanical, thermal equilibrium is not satisfied, there can exist no thermodynamical equilibrium. In essence, thus, there is never an equilibrium, and everything is 'changing', or 'animated'. No life force needs to be postulated here, in contrast with spiritualist theories.

Something, no doubt will upset Libb in his categorization. He has used the animate-inanimate distinction to replace the biological-non-biological distinction, he must realize that his categorization is useless.

The thing about DNA

David: *also one thing I found in common with all 'living' things is DNA. while I am of the opinion that, in other conditions, some other molecule will represent the 'code' for the cel division- it is noticeable that DNA is, in some way, a 'life molecule'*

For instance, in alien civilizations, some other molecule might exist that has roughly the same function as DNA.

Jeff: *This is another very good question! What is DNA for? It is for passing along information (chemical evolution) between some really complicated physical systems (pear tree, elephant, etc.). Here you have to ask yourself, is DNA the cut-off criteria for determining an animate from an inanimate system?*
So a rock has no DNA, and must be inanimate, whereas a squirrel has DNA and is animate? I think (and I might be very wrong!) that DNA is a fancy molecule which helps in the evolution of really advanced (evolution-wise) chemical systems. But there are a huge number of other big macromolecules which also help other complicated evolved systems, its just that they either play a smaller role over a shorter amount of time, or are found in both DNA and DNA-free biological systems. So DNA in my opinion is currently the cream of the crop in terms of long-lasting macromolecules, but it still does not mean that there is a strict difference between the animate and inanimate in the larger scheme of things. But hey, I may be wrong! You should look into this more, it is something I never really considered.

David: *DNA vs no-DNA is still a difference, but it depends on which level we are talking.*

Is the subject of DNA a solution to the origin of 'life'? Yes and no. Life is commonly defined as having a 'soul', at least by Libb - a concept which stems from religion, and dualistic visions like Descartes, that our mind can detach itself from matter in an immaterial world. There is no life force out there that detaches us from the laws of physics or chemistry. I'm not even necessarily saying there is a difference between living organisms and the other, besides DNA. it's just that what the common view of 'living organisms' corresponds perfectly to 'having dna'.

The discussion went further about interesting books Jeff recommended me (one of which seemed very interesting -Wicken's "Evolution, Thermodynamics, and Information") but Jeff knew another interesting thing to tell me:

Jeff: *Also, regarding your previous question about entropy: From a statistical mechanics perspective, there are two distinct sources of entropy. The first is "thermal" which has to do with the number of states in which the energy can be randomized. The second is "configurational" and has to do with the number of possible structural arrangements (you can think of this as spatial arrangements) of its components.*

Interesting facts... However the conversation takes the turn to DNA again.

Jeff: *"DNA is after all, a pretty darn important molecule, but is it a real, physical threshold for something so fundamental?"*

David: *If you assume there to be some fundamental difference between biological and non-biological, then you are postulating a difference between the two - besides DNA. I am however only stating that there is only one meaningful difference between biological and non-biological that is everywhere- namely genes. There could be others, but that would be invoking some spiritual force for most of the folks.*

Erythrocytes: semi-biological organisms inside a biological organism

Jeff: *think we have to reasonably ask what is the difference between biological and non-biological in order to get off on a sound footing. For me, there is no clear difference. DNA is a powerful molecule, but it can exist outside of an organism, and be synthesized in a lab just like many other* macromolecules. Where then does the DNA come in as the one common characteristic between biological and non-biological?

Also there are some known cells, like erythrocytes which have no DNA, but are for you and I a very important part of our "biological system." There are trace amounts of some elements such as Vanadium, which are normally considered non-biological (not an organic compound) but which we require. This is why I think we should be careful in examining the DNA = life discussion. In jumping to conclusions, it is often easy to sprain one's ankle.

David: *erythrocytes are not a whole organism. it can be synthesized of course, but that does not deny that what we refer to as life has one common characteristic.*

if dna is not inside some cell, then it cannot perform its function of cell division, which is why nobody would refer to that as life.

David: *it seems unnecessary to make a new name for something that already has a clear definition : biological organisms ='having DNA as coding for its proteins'.*
whether you want to call it alive or not, for the sake of practicality we might as well choose an existing name.

Jeff:

If you agree that an erythrocyte is a cell, but has no DNA, then you must come to the question "are cells alive?"
If your answer is yes, then DNA cannot be seen as the criteria for living systems.
If your answer is no, then you have to ask the next question, "is DNA (are molecules) alive?"
The answer to that question, I believe can be found here in Hmolpedia with little effort.

David: *then of course we would have to say cells, although most of them are, are not necessarily 'biological'. anyway some minor exceptions like the erythrocites, or perhaps a virus, which has only RNA (perhaps we should change the definition to RNA then), do not negate how simple this solution is to denominate something that is clearly of a lot of practical value.*

David: *hmm, that's a jump there. If you define 'life' as organisms that have DNA coding for its proteins then your 'next question' is not really relevant is it*

David: *you make the mistake that I should refer to some 'life force' inherent in DNA - where I only say that biology can easily be summarized as the class that has genes coding for its proteins.*

Jeff: Yes, it does negate it. Taking liberties such as "well, most of them fit into this catagory" and "well, for the others we can use RNA instead" is not science. You may as well say that 1+1 is not always 2.

RE: *"If you define 'life' as organisms that have DNA coding for its proteins then your 'next question' is not really relevant is it"*
Yes it is. You first have to ask the question "are cells alive?" If your answer is no, then holding your belief that something must be alive, you can only go up in scale "organs are alive, but cells are not" or down in scale, "molecules such as DNA are alive, but cells are not."

Molecules are not alive. That is a misinterpretation of my definition. I have defined a living organism as 'having DNA coding for its proteins'. Does the DNA molecule have DNA coding for its proteins? Or even, should my toe contain toes, so that my feet can be rightfully called feet?

Jeff: *The statement "biology can easily be summarized" frightens me. Please go outdoors, collect a few simple leaves and dissect them. What you will find is nothing which can be simply summarized.*

David : *I meant: 'what is unique to biology is ...'*

David: *one can say that $x^n + y^n$ is not always z^n. is that not science? –*

added note: I did not mean that we should use RNA only for the others, but for all of them- or for none- but this will become clear in my next post.

David: *'Yes it is. You first have to ask the question "are cells alive?" If your answer is no, then holding your belief that something must be alive, you can only go up in scale "organs are alive, but cells are not" or down in scale, "molecules such as DNA are alive, but cells are not.'*

see, if I define 'biology' simply as 'having genes coding for its proteins' then there must be no question asked if the cell [without DNA] is alive. it might be a part of a biological organism . what the answer is on the erythrocyte doesn't really matter does it. if I , starting from now, define life as having DNA coding for its proteins, then you will find few exceptions in what we call life. if there are exceptions, why not cut those exceptions, if the only common characteristic of life is 'having gene coding for its proteins'. so, for example, why not say erythrocites are a non-biological part of my body- as for instance when I put a chip in my brain. and why not say, a virus is not biological- or we can broaden our definition by saying 'all that has RNA'

Well, I want to add something: if we define a 'biological organism' as 'a structure composed of cells, whose proteins are coded for by DNA or RNA', then it becomes clear that this is a very solid definition. We don't even need to say that there is a non-biological part, since we refer to the whole organism at once. But perhaps we can make a distinction between these 'biological parts of a biological organism' and 'non-biological parts of a biological organism'. Better however is to say that erythrocytes are semi-biological, since each of them consists of a cell – which is one of the two criteria. Bones and teeth are special cases as well: some parts of them (enamel crystals) do not contain cells – though clearly their structure is initially generated by cells. It is these two cases that are relatively intact for biological degradation – skeletons can remain intact for quite a while.

Then Libb comes in and asks me to tell me when the 'magical' event happened in the evolutionary time scale that 'life' originated.

David: *well, I would say biology started when cells (having DNA) began to form. if we don't want to call it life for religious connotations that's fine with me. just to mention there is one distinction between what is commonly referred to as life and what is commonly referred to as not life, and that is the possession of genes. This cannot be denied. I do not mention free will, soul, spirit anywhere.*

David: *also, I don't think it to be a right approach to answer the question what is life/biology/organisms using only one principle. for instance, many characteristics are exclusively biological, yet they might not apply to all. for instance, genes that code for proteins, are clearly pretty congruent with the term biology. other instance: cells. just because there are some sub-distinctions does not mean we can deny that bio/whatever- ology is different from a rock, a table, a chair, glass...*

Surely there must have been some beginning of the existence of DNA, or more generally genes (DNA or RNA), so clearly this is where 'life' as we commonly know it begins. However, the term life may hold to many spiritual connotations for all the abuse it has known in the past. Two terms can be proposed: biology or organisms. Or perhaps the term 'animate' vs inanimate can also be proposed. In that case, however, the gradual nature of the term animate as in our discussion above - where 'much animation' refers to strongly in disequilibrium, and 'few animation' refers to almost in equilibrium – would vanish. Perhaps biology still is easiest because it specifies a all-or-nothing contrast.

Also, suppose we did not have an unanimous answer what principle there is behind 'life'. Does this mean that 'life' does not exist? It could easily be that we are simply too dumb to find it.

While not believing there to be a single 'principle of life', or a 'life force', we can of course not deny that biology- if that would still be a correct term- has its own distinct properties: DNA, RNA, cells, organs are functional parts that are not seen in rocks, lamps, etc. Of all these I believe DNA, and having cells, to be the best demarcation for biology – although there are exceptions such as erythrocytes and viruses which I at least would consider biological. From now one we term these two *semi-biological*. The mineral/crystal parts of teeth and bones have no DNA, nor cells and thus I do not see them as biological. Yet they are embedded in a larger biological whole, since they indirectly (by means of ameloblasts) rely on DNA to code for their proteins.

What to make of the strange example of the erythrocyte? It has a special status since it is considered a cell – but has no genes (DNA or RNA). This special status also allows it not to get infected by viruses. Perhaps there is some linkage with the story of both of these 'monstrosities'.

Perhaps, and I am making a wild hypothesis here, some viruses were once the DNA or RNA inside the erythrocytes, but somehow managed to get outside. The virus then found hosts, and the erythrocyte population didn't really need genes for cell division and repair mechanisms, since bone marrow readily produced new erythrocytes on a continual basis. This, I believe, is a possible origin of these semi-biological organisms. I name this the erythrocytal-virus escape theory.

If viruses originate from endogenous cells, why then are they usually harmful? The viruses can originate from other animal types with erythrocytes, and so only in certain types of animals will these viruses be malignant. Or perhaps viruses were once harmful against certain types of disease processes, thus making them adaptive for the body.

Viruses cannot infect erythrocytes, so this can be seen as either confirming evidence for this hypothesis (if viruses stem from erythrocytes, then it would be horrible if they infected them), or as falsifying evidence (erythrocytes evolved to be without DNA so that they couldn't be affected by viral infections; viruses were never inside and they were unwanted from the beginning).

Why my toe should not contain my feet and why categories should exist

Why Libb has problem with accepting a difference between biology and non-biology is that he assumes that the term life implies that you imply that all individual molecules are alive. That's not really an issue if you define a biological organism as 'having genes as coding for its proteins', since the only necessary thing is that at least some of it cells have genes, coding for some of its proteins. We cannot say that 'there needs to be a DNA molecule inside the DNA molecule so the DNA molecule is alive' – we are defining a biological organism as a whole. Containing DNA molecules which synthesize proteins for the body is enough!

A question that may arise is 'is there a rigid definition of a cell?' Yes, it seems to be the case. The cell membrane, intracellular liquid, nucleus, lysosomes, mitochondria, microtubule are all characteristics that are present in all cells – excuse me if there is one exception in a billion – and these characteristics in turn are very well defined as well. For example (see Wikipedia): The cell membrane consists of the lipid bilayer with embedded proteins. Cell membranes are involved in a variety of cellular processes such as cell adhesion, ion conductivity and cell signaling and serve as the attachment surface for several extracellular structures, including the cell wall, glycocalyx, and intracellular cytoskeleton.

The argument on the status of the word biology then proceeds, with Libb.

Libb: Your argument "I would say biology started when cells (having DNA) began to form. If we don't want to call it life for religious connotations that's fine with me." Your argument is circular, you juxtapose two synonyms (life and bio-) to make your argument sound presentable.
Simply because a molecular structure aggregates a strand of ribonucleic acid into its composition is nothing in itself special other than an added molecular property of structural repetition in product formation.
The reaction that brought about this structural repetition synthesis is the more important factor and one driven by free energy, which is where you need to re-focus your investigation. Certainly, you are within the bounds of sensible science to label yourself as a RNA-centric molecule (or off the class of RNA centric molecules), but not a "living molecule", because in order to do so you would need to overthrow the entire history and science of chemistry by your splendid proof that molecules are alive and that the hydrogen atom alive. This will never be done. Hence new language is need. The old language of "life", bio-, living, etc., are fine for the layperson, in the same sense that we speak about the sun "rising", but in the strict scientific sense of the matter, the sun does not raise (an Egyptian mythology theory) any more so than are humans alive (an Egyptian mythology theory).

RNA-centric molecule might be a fine definition as well – if the human and other biological species can be considered as a molecule, that is.

David:
well, life sounds a bit more religious than does bio, no circular reasoning need to be involved. 'Simply because a molecular structure aggregates a strand of ribonucleic acid into its composition is nothing in itself special other than an added molecular property of structural repetition in product formation.'
do I say that then? I'm saying that what is defined as biology these days perfectly corresponds with organisms that have cells and dna. if you look at things that do not correspond to what is called biology such as a chair, a lamp, you will find that they do not contain dna or cells (both cells and dna are very strictly defined).
and why I say life sounds a bit religious is because it recalls of soul (and thus free will, a 'life force' if you will), and if for one way or another bio would recall of the same, then we should leave that term as well. But since biology is a science by now, that has a good definition not relating to free will, it would be somewhat unnecessary to change the definition of a clearly defined subject. life on the other hand is a broader term that specifies much more such as all things, daily life, biology, and has connotations with a 'life force'.

"overthrow the entire history and science of chemistry by your splendid proof that molecules are alive and that the hydrogen atom alive. This will never be done"
you're not applying my definition of life then are you? do dna-molecules have dna to make their proteins, and do they consist of one or more cell? do atoms have dna to make their proteins, and do they consist of one or more cell? no.

if I say my feet contain a property, should that property also extend to my toes? following your reasoning, one could say: if my foot has toes, then so should my toes have toes. Analogously, if my body contains DNA and cells, then should my atoms contain DNA and cells?

Quite clearly, if some property is defined such that the minimum scale is that of cells, then clearly, there is no need to prove that something smaller than this minimum scale of the definition should have cells!

So in essence, at this point I think I have a clear demarcation of biological vs non-biological. Why Libb and I did not agree was mostly because he didn't apply my definition. This illustrates again how crucial a clear definition is. In one sense I have agreed with Libb that everything follows molecular forces, biology being no exception. In the other sense, it is clear that biology differs from non-biology on a somewhat larger scale, because biology contains cells and DNA.

Perhaps we could term this mistake the 'chemists mistake': while molecules will remain molecules, it is clear that on a larger scale, there exist clear patterns distinguishing biology from non-biology- this does not imply that its molecules are subject to different laws, it just means that the one has some properties that the other doesn't on that scale. However, even chemically, we could see patterns: for instance phospholipids on cell membranes, the chemical structure of DNA and ATP vs no phospholipids, not the exact structure of DNA, nor ATP. ATP's formula, for instance, being $C10H16N5O13P3$. But still, nowhere any exception to chemical laws of course. On the smaller scale, say the quantum scale, we are not saying anything here – but neither are my contenders.

My stance can best be summarized as: 'Having a sound definition is very important to make a concept defunct or not. I'm pro defunct life theory in the sense that there is no 'life force' that seperates biology from non-biology. However, if we adhere to a philosophy where we cannot make distinct categories, just because they are all comprised of a smaller category of atoms, then we have no categories at all, except for atoms. One would end up only recognizing that the only category left is strings, or even further down the scale. Saying life is defunct only is useful in the sense that it emphasises that we are determined by chemical forces. Even if we look at chemical structures, we would find patterns in biological organisms not present in non-biological 'things', for instance phospholipid membrane, ATP, DNA are chemical patterns very common in biology. Does that mean they are not subject to the laws of chemistry? Does that mean they have a soul?'

After this, Jeff agrees with most of what I was saying.

It surprises me that a very well-defined concept such as biology is under attack, whereas concepts such as love and evil are not strange to Libb, even in his scientific works. How do we define love? Can we define it chemically? We can: for instance, chemical response relating to sexual contacts. Of course this is only one kind of love; but it is the one that gets most attention in Libb's vision of the elective attractions. But, if we would follow Libb's strange philosophy that 'if something can be explained by chemistry then we cannot make any categories on higher levels' , then Libb would end up saying 'love' is a defunct concept as well since love is also chemically determined.

The conversation keeps going as Libb cannot properly use my definition, and by the way he has removed my stance on the issue . At the moment, I thought I had obeyed all the rules of editing on the Hmolpedia: third person, reference, objective standpoint. Anyway, here's what happened:

Libb: *reverted your add. Please don't add your personal theories to articles.*

Your:

Life = Cell + DNA

model is easily shown to be defunct according to the cell as molecule model:

At what point (second), for example, in the cell molecule formation reaction

A + B --> Cell

did did the entity become "alive"? The hydrogen, carbon, nitrogen, etc., atom reactants certainly are not alive. But according to you, at some point along the line, this DNA-containing molecular cell-shaped geometry became "alive".

David: *you're obviously confusing life and free will. I don't say there is a fundamental distinction between a molecule inside a biological organism and other molecules. if we define biology as 1. having a cell as its smallest unit 2. having dna to code for proteins, then clearly if you say I should prove hydrogen is alive, then you are not applying the definition of biology are you? should I prove that hydrogen contains one or more cells? should I prove that hydrogen has Dna coding for its proteins?*

Quite clearly, *nothing* on a scale lower than the cell can be considered as biological in this definition.

David: *well, about the add: there are other personal theories there? If I use third person and reference to its source it should be ok normally?*

Indeed, the whole page was full with personal theories. From Tesla to the common blogger. Note that Libb often himself comments on stances, without any reference who the author is or what article the article came from – i.e. giving his own personal theory, without labeling it as such. For instance, "Stamatopoulos' bipartisan siding here is a good example of the newcomer's reaction to the issue" is one way Libb intervenes with his own opinions on the matter, and with a grain of ridicule to bias the reader. Perhaps it is more a question *whose* personal theories are allowed to be represented by *whom*.

Other people have had trouble with this as well. A search on Wikipedia gives quite a lot of complaining contributors, mainly in terms of misrepresentation of external articles (as with my squidoo article that will follow a bit later) or in terms of pushing his own theory.

Anyway, the conversation then proceeds:

Libb: *The life equals cell or structure with boundary (Margulis, 1995) + DNA (Crick, 1953) models of life are listed here:*

http://www.eoht.info/page/Life

There's dozens if not hundreds of "life equals" theories. The new hmolscience view is as follows (Carolyn Porco, 2006):

"Our reconstruction of the history of the early solar system and the events that led to the origin of the earth informs us that around 3.5 billion years ago or earlier a complicated chain of chemical events

occurred to evolve animated molecular structures from inanimate atoms."

All there is are atoms and movement; there is no "life". To repeat, in the famous 1925 words of Tesla "there is no thing endowed with life."

This view solves many hindering issues, such as the debate as to when "life" was thought to start in regards to reproduction and abortion issues. Just as you can't find or pinpoint a so-called start second to life in the evolution point of view so to can't you find a start second to life in the conception process, which technically is a double displacement reaction:

http://www.eoht.info/page/Human+reproduction+reaction

Try to let all of this soak into your head for some time.

RE: "well, about the add: there are other personal theories there? If I use third person and reference to its source it should be ok normally?", typing up some random objection comment in the threads and then citing the thread as a reference is only rarely done here. The main focus of Hmolpedia is to read books and articles, related to the hmolsciences, and then write up the important terms, theories, equations, and people, of those books and articles so that we can collectively study the various historical paths at attempted solution to the big problems of the universe.

Again Libb cannot see that there *is* a pinpoint to biology, namely the origin of single celled organisms. Why birth and death are confusing terms is that, indeed they do refer to a sudden transition. I prefer to talk about 'slow biological genesis', or simply 'cell division' (dividing of the cells in the uterus), and 'slow biological degradation' (when being called 'dead' there is a long process of biological degradation, however, the events that precipitate this degradation *are* sudden) – which reflect that neither of these processes is sudden, as the terms birth and death do imply, reminding us of the life force of the soul assumed by religion – imagine those cartoons where ghostlike souls fly out of the body of the deceased. In the case of birth of course, we may define it as well as 'exit uterus', which *is* a sudden transformation.

David:

"Try to let all of this soak into your head for some time."

I believe I already did, but I will ponder further on the question. However, I still see no reason why one cannot categorize a chemically determined thing as 'biological' using my definition- it is in my opinion quite similar to categorizing a quantummechanically determined molecule correctly as a molecule- as I also don't see anywhere in the defunct table, that this definition has been proven false - for instance even with DNA alone as a criterion there is no comment.

Again I speak here of the weird assumption that things that are determined on processes on lower levels cannot be described on higher levels. This would amount to the faulty conclusion, when taking the lowest level as the strings in string theory, that *everything* is a defunct term, except for the terms for these low level strings.

About abortion: the criterion for a moral use of abortion is clearly not whether something is biological – since one -, two- or four-celled organisms for instance will be aborted without any doubt on moral issues. What seems to be the criterion for most people, is whether the being has emotions or consciousness. Again, Libb's comment may reflect that his definition of life is quite similar to his definition of consciousness.

Generally, I get the impression that Libb sees life as 1. A supposed physical force present in the atoms of 'living things', but not in those of objects 2. A force that accounts for the supposed consciousness. 3. A soul.

Previous attempts at definitions of life

There was a table of the attempts to define life which according to Libb all failed. I noted here that he did not give an explanation with regard to DNA. The others were for instance *That which moves?* Molecules move and are not considered alive? *That which evolves?* The universe evolves, yet is not considered alive. *That which is intelligent?* Computers are intelligent. *That which can reproduce?* Computer viruses can reproduce.

Libb takes one out of many aspects and then considers them each separately. If one would say : a biological organism is that which has 1.cells 2. Genes as hereditary component 3. Genetic constitution gradually changes over the course of history of recombination and mutations ('genetic evolution', not just 'change'). 4. Systems that have these 3 properties can have the ability to reproduce, and optionally have a relatively high amount of intelligence , then there is no problem with the definition of biology, nor genetic evolution, is there?

Somewhat inappropriate to take one characteristic separately, for instance intelligence, or reproduction, without checking satisfaction of the other characteristics (cells and genes, the very core characteristics), to hence conclude that biology is not well-defined. Intelligence and reproduction are not 'core characteristics' of biology, since there are clearly non-biological species that reproduce, or are intelligent, and on the other hand, there are biological species that are not intelligent – depending on what threshold is defined for intelligent of course. The 'core characteristics' of biology are the two criteria I have mentioned. 'Marginal characteristics' include organs, intelligence, and reproduction. These characteristics, however frequent in biological organisms – reproduction actually is, as far as I know, present in every organism – these characteristics can be found in non-biological 'things' as well: computer viruses can reproduce indeed; computers and AIs indeed are intelligent ; even a car has organs in the sense that there are elements useful for a particular function, to maintain global functioning - the difference in biological organs of course is 'to maintain global function of a biological organism'. So these three are in my opinion not exclusive enough to biology, to define them as core characteristics.

Furthermore, it doesn't arise to Libb, to make subcategories in biological organisms. For instance, one could define biological organisms that are intelligent as 'intelligent biological organisms'. One could define organisms with no cells or with no DNA as 'semi-biological'. Simply having one or two counterexamples does not render a whole category obsolete! Rather, it indicates that some elements are not in their place in their current categorization.

We were saying that there was no counterargument provided against 'has genes coding for its proteins' nor against 'has cells', the two core characteristics.

David: *Is love a defunct term? I ask this question, because, similarly to your argument that, if something cannot be extremely precisely be defined, then the concept is useless, has no function. And even if you ask something to be precisely defined I have done that, and if there are 2 counterexamples in a billion, then we can easily say that those things are NOT biological. the term biology would still remain useful.*

Obviously, there is a large consensus of what constitutes life. Otherwise, I wouldn't get responses like 'although this feature is absent, it is still recognized as biological', or '…yet the bacteria is not considered 'alive' ' Furthermore, if we have not found a clear definition of life, that does not mean it doesn't exist. We may be too dumb to define it; we may not know all the factors. Analogous with the definition of continuity in mathematics, which was only then rigorously defined when Weierstrass came up with the epsilon-delta definition, it is conceivable that there is a large temporal gap between the origin of a term and a solid definition. Most terms start of somewhat fluid, vague, only to later be anchored in a scientific definition.

At last, we seem to be at a point of agreeing:

Libb: *The issue is not with the superficial term "biology" but with its root "life". Long ago, Schrodinger famously attempted to define life in chemical and physics terms; concluding in the end, following attack from his physics colleagues, that the discussion must be turned to free energy. When you do in fact turn the discussion to free energy, which amounts to the reading of every page of Hmolpedia (or in my case the writing of ever page of Hmolpedia), the definition of life falls apart and it takes some years to see this, hence the suggestion "let it soak for some time". If it is something that cannot be defined by physics and chemistry it does not exist. Life is one of these somethings; a model passed on to us from ancient times that each day the sun is "born" (given life) each day and "dies" (life taken away) each night. This model, however, does not corroborate with the periodic table and the reality of chemical reactions.*

David: with this I agree. but then life is defined as the magical life force that has not been found, and will never be found. I'm only arguing that biology, as the category of organisms that comprise 1 or more [cells] and have dna coding for its proteins, is a term still very usable, and still very practical, and pretty exact. So, if we would define life as a force - present in animals, but not in things, then the term should be abandoned - but again, biology should be there to stay. Man, I should be studying :D. perhaps I can some time on squidoo post my stance, if it is somehow relevant for a page. I will let you know when I have done this, then you can still choose whether it adds to one of the pages.

Libb: *Yes, posting your stance on Squidoo is the right idea. Here's a relevant quote, for you to think about in the mean time:*

"How, therefore, we must ask, is it possible for us to distinguish the living from the lifeless if we can describe both conceptually by the motion of inorganic corpuscles?"

—Karl Pearson, The Grammar of Science (1900)

With this beautiful quote our discussion has ended, and I decided to summarize my standpoint on squidoo to be referenced in one of the pages. It should be clear that the quote is not contradictory to my views, where indeed we are mechanically and chemically determined, where the term life - as in a life force, a soul- is fiction, yet where the term biology still holds meaning, namely: having, or consisting of 1 or more cell, and having genes code for proteins.

Applying definitions

My squidoo was as follows.

Personal view

Where Libb emphasizes the fact that the living, just as the 'dead' things like a table, are determined by exactly the same laws, I argue that that there are subtle distinctions which need to be made to correctly declare defunct. In other words, we need a clear definition before we say that a word has no sensible meaning in scientific terms.

Therefore, the term life should be separated from the term biology. Where the term life holds many connotations, such as a soul, and not-determined, having free will, the term biology does not. Of course, life holds many other connotations, but we discuss here as is relevant to the debate of 'life, a defunct concept'. So, a solid definition of life would be: a force that moves biological organisms and not 'things'. A very exact definition of a biological organism would be: 1. has, or consists of one or more cell ; 2. has genes (DNA or RNA) coding for its proteins.

So the term life implies some 'animus', a 'spirit', reminding us of the dualism debate stemming from Descartes. This life force - even should religious people insist it does- will of course never be found. We cannot find any such force that exists in animals, but not in things. We are obviously all object to the same chemical laws of free energy.

Yet, the term biology to me, makes much sense. Not only does biology have a very clear definition, with very few exceptions, it is clear that in terms of categorization, its practicality is high: it is not because biological organisms are dictated by the laws of the small molecular, that we should not recognize its existence, for this would imply that we do not recognize chairs being different from animals, but even chairs being different from tables, since they also are determined by the same laws. Also, I have been told that I should prove that atoms are alive or biological, if I adhere to the vision that the concept of biology is useful. But that results from not applying the definition I have given above: if one applies that definition it is clear that atoms need not be biological, since they are elements OF the molecule of DNA, and OF the cell. So the atom then, cannot be biological, since it is not the level that has relevance for the definition- an atom cannot contain a cell, nor can it contain DNA.

Proving that the atoms are alive is unnecessary too, since the definition of life implies an unknown force that is present in biological organisms, but not in things - which the definition of biology has not postulated.

So, is life a defunct term? Yes. There is no force present in biological organisms that is not present at 'death'; there is no force which makes biological organisms less determined by external forces. Is biology a defunct term? No. Its definition is clear and scientific, its practicality is high.

The term biology can also explain the paradox of death. While there is no distinction, in terms of forces, we can say, that at the biological level, for instance, the cells have stopped dividing - the telomeres were too shortened; or the hart has stopped ticking, disabling blood flow to cells.

Libb then adds my view, totally misrepresenting what I said. Let us see.

Libb:

In a Squidoo article entitled "Life, a defunct concept?", shown adjacent, Bossens summarized his view that "life" should be redefined as "a force that moves biological organisms", where biological organism strictly means an entity that "has or consists of one or more cells and has genes (DNA or RNA) for coding its proteins."

Bossens' logic, however, is error-ridden on a number of points. Firstly, Bossens' definition is recursive amounting to "life is a force that moves life", in short, which is meaningless. Bossens seems to think that by mixing together Greek (bio-) and English (life) synonyms of the same word in once sentence that he has achieved something? Secondly, an easy disproof his his life definition is the example of a resin fossilized coccoid cyanobacteria from 3.5 billion years ago, such as depicted on the evolution timeline, which clearly (a) has one or more cells and (b) had DNA, but does not, however, in the colloquial sense, seem to be "alive", as Bossens' definition would entail. Thirdly, he states "biology has a very clear definition". This could not be farther from the truth. Biology—by world-over agreed upon definition—is "the study of life" and as Bossens answers later in his article "So, is life a defunct term? Yes." This amounts to Bossens' new 2012 definition of "biology" as "the study of a defunct term".

Hence, the article is recursive and circular all in the name of "practicality", which seems to be his objection, i.e. "biology is a practical term", thus we should keep. Bossens here would be wise to heed the famous words of Aristotle: "Plato is my friend, but truth my greater friend." If there is no truth in a term it should not be kept.

Libb refers in all his writings that life should be a property present in atoms. This however would assume that life is an atomic force or property (and not just a thing that has cells, and genes coding for its proteins, where no separate atomic force is postulated). This, life as an atomic force, is clearly not how *anyone* defines the study object of *biology*, which Libb thinks equates with the study of life, although he is not aware of it – both what he implies (when he says the defunct life theory opposes immateriality, spirituality, and the 'atoms are alive'-hypothesis) and what anyone else would define as biology. I have, maybe, defined biology different than others – who would term it 'the study of life'. However, it corresponds neatly to the category everybody envisions as biological organisms – or what is studied in biology classes.

It should be clear that Libb never even gives a definition of what life is. So what concept is he debunking? One of the problems is that Libb uses defunct in a double meaning (again a problem of definitions): sometimes he uses it to refer to not-existing or debunked by science; other times he uses it to say that there is no concrete definition, for instance by arguing 'what is life? Is it that which evolves? No, the universe too evolves'. Now I have clearly defined biology and life, the only thing left that may or may not show the defunct status, is to show that they are or are not debunked by science. Is biology, satisfying the two criteria, debunked by science? No. Is life, according to my definition, debunked by science? Yes.

For those who insist asking why I should make the distinction, well I discuss this later, but I give one hint: a response to those who refer to life as a physical force.

It should be noted that biology and life are already two different terms, even if their etymologies track down to the same root: life is used in many contexts to mean social activities, apparently also a soul, whereas biology is mainly used in science.

So I started a Hmolpedia thread called 'misrepresentation' :

David: *"Secondly, an easy disproof his his life definition is the example of a resin fossilized coccoid cyanobacteria from 3.5 billion years ago, such as depicted on the evolution timeline, which clearly (a) has one or more cells and (b) had DNA, but does not, however, in the colloquial sense, seem to be "alive", as Bossens' definition would entail"*
you have apparently not read my definition of 'life' is, and what my definition of 'biology' is. you're misrepresenting what I said in many ways.

Note that my definition not at all implies that any biological organism is alive, since I have said that 'life' is a force that does not exist, whereas biology… (see A and B). The cyanobacteria is clearly biological and *not* alive, at least according to the Bossens definition. With regard to fossils, very few of them will actually contain cells or DNA. If they do – in which case Libb should have used the present tense 'has DNA' - then that may seem as if it discredits the fact that the two criteria are equivalent with a category that most would call alive. However, this still is incorrect, since it relies on a loose interpretation of one the two criteria: 'has DNA' is incomplete. In the squidoo article, I clearly say that one of the criteria is 'has DNA coding for its proteins', not just 'has DNA'. In fact, even I was incomplete in the squidoo and I should have mentioned that DNA should also be renewing cells. Both of these functions are not being performed by DNA in fossils.

> David: "Firstly, Bossens' definition is recursive amounting to "life is a force that moves life", in short, which is meaningless"
> again, misrepresentation. not only have you agreed that the term biology and life are different, in your last post , but also I have not said 'life is a force that moves life'; i have said that the 'life force' (defined as 'life') which will never be found, would entail a force that there is a force present in biological organisms but not in non-biological organisms- however, I resent this life force. that does not mean that biology cannot be categorized as different from non-biology, if we use the definition given.
> "Yes. This amounts to Bossens' new 2012 definition of "biology" as "the study of a defunct term".
> Not really, since I have clearly made the point that if we use the term life, that biology is not the study of 'life', but the study of organisms that have 1. and 2. (you know by now what 1 and 2 are).
> it seems you point to my 'logical errors', which are rather a result of you not applying the definitions I have given for the two terms. And even if you say that one of those definitions is not according to what you or someone else believes, see them as a way of letting you know my exact stance. if you specify definitions before you talk, then you can talk.

Notice how Libb intervenes to change the definition. Libb makes the following reasoning: "suppose we define life and biology according to the Bossens definition. Then, if I switch the definition of biology to 'the study of life' (which is not the Bossens definition), then I end up that 'life is the study of a defunct concept', according to the Bossens definition." Big surprise, if you switch definitions!

Even if my definition would not conform to what others think it is, it should still, to a sensible reader, be clear that I am using this definition to make clear what I think exists ('biology' via the new definition) and what does not exist ('life' via the new definition).

> you have also not even mentioned my definition of life, and thereby you let the reader believe I make errors by simply not using my definitions - as we see in your sentence 'life is a force that moves life' something I never said, and nothing what I said implied.
>
> also, I don't know if it is ok in this wiki, but you keep adding your own opinion of the matter without referencing to any external document, nor referring to that it is your stance.
>
> the only error in reasoning lies in your shifting from my definition towards the definition of others - by assuming life and biology are defined as the same- so that you can justify saying I reason circularly.

Even after clearly making my point *again*, Libb refuses to apply my definitions correctly. Note also that, without Libb, I wouldn't even *have* to make this distinction between life and biology. Over the course of many explorations of Hmolpedia pages, I was confronted by his view of life as defunct- stressing the fact that there is no physical or chemical force separating life from not-life – also often it was stressed that there is no spirituality. Initially, I was convinced by his arguments that 'atoms are not alive'. After many philosophical investigations, I stumbled upon the fact that 'has DNA' was a very plausible candidate for the distinction between life and not-life. Therefore, the distinction had to be made between this life force present in atoms *he* implied as the definition of life (A) and, on the other hand, the fact that the things we used to call 'living' (B) is a clear cut category- on which everybody, except a few in a billion counterexamples agrees on. (Note that the definition of 'entropy' is under debate as well, as seen in the Mortiarty-Thims debate – let's not suggest that entropy does not exist!) So, the first, I would now refer to as 'life', and the second I would now refer to as 'biology' or 'biological organism' – in the course of the conversation I decided 'has cells' is another uniquely biological trait.

The argument that practicality should not prevail 'truth' is not really truth itself: not only does truth imply practicality (science that doesn't work is most likely false), it is as well the case that 'truth' here is defined as 'something that is determined by lower-level laws cannot be classified as anything different than from a bottom-up perspective.' If we know that a lamp follows the same laws as an animal, albeit that they have a different configuration of components and appearance (indicative of different wave lengths, mass, height, compressibility, movement, sound, so not *just* appearance), should we say that we cannot categorize the first as lamp, and the second as an animal? In other words, chemical determinism, or even more-down-the-scale-determinism, in no way implies that we cannot categorize things on a higher level, given several clear characteristics. Some chemists like Libb however, argue that we should call all things a molecule; that's fine with me if that's one way to approach the problem, but I'm not going to say that this implies 'biology does not exist', nor am I able to give one common chemical formula for 'all organisms with genes coding for proteins, and having cells' – the same can be said for computers, who clearly all have different compositions, yet every person would agree, given some characteristics – mainly at higher levels than chemical (function, form, appearance, etc.) – that they are indeed the category of computers. The same holds for biology, there are clear *patterns*, as well on the higher level, as on the chemical level. With 'practicality' I thus clearly refer to something that *is* truth: everyone can agree that what is commonly referred to as 'biology' is correspondent to the criteria I proposed. Two counterexamples exist in a billion: erythrocytes have no RNA or DNA; viruses have no cell- although the discussion is heated whether viruses really are considered biological organisms. But as I said, we can easily categorize erythrocytes and viruses as *semi-biological*, also referring to my hypothesis that they share a common origin : viruses perhaps were once the DNA/RNA that was inside the erythrocytes. Furthermore, an erythrocyte only exists within a biological body. As implied by the definition of biology I proposed, not all parts of the biological system must be biological. For instance, bones are mostly non-biological - since a part of it is not cellular in nature. Still, the surrounding whole contains predominantly cellular matter, so the being is called biological.

Even peripheral cases such as bone mineral and enamel crystals their existence is fully determined by the presence of cells and DNA – hence the larger system is without a doubt biological.

Libb in another post told this:

This is your definition:

"Life is a force that moves biological organisms."

which amounts to:

"Life is a force that move life organisms."

which in shortened form is:

"Life is a force that moves life."

Spend some time reading about circular definitions:

http://en.wikipedia.org/wiki/Circular_definition

Also instead of arguing with me spend some time trying to disprove Tesla: "There is no thing endowed with life."

Clearly, Libb cannot apply my two definitions consistently, but rather chooses to replace my definition of biology with his definition of biology (the study of life). I then replied that he must look what I wrote in the squidoo, to make my stance more clear.

I sometimes get the comment that I do not use the definition or the terminology that others use. If that is really true, then seperating the two - 'life' and 'biology' - still holds much meaning: the one doesn't exist, and the other does. What term we use is not that big of a deal, but, since we already use the term biology for a category that corresponds neatly with what is actually phenomenologically referred to as biology - go over all the examples, and you will see that biology is precisely this category I have defined with the two criteria.
But then I get comments like these : "Secondly, an easy disproof his his life definition is the example of a resin fossilized coccoid cyanobacteria from 3.5 billion years ago, such as depicted on the evolution timeline, which clearly (a) has one or more cells and (b) had DNA, but does not, however, in the colloquial sense, seem to be "alive", as Bossens' definition would entail"
I have nowhere alleged that cyanobacteria are 'alive' - since I consider NOTHING to be alive - recall my definition of a seperate physical force, present in biological organisms, but not in non-biological organisms.
For those arguing that a phenomenological approach is not tenable, note 1. you can then not make comments like 'they seem to be alive', since this clearly refers to a phenomenonological approach. 2. that DNA and cells are not only phenomenologically present, they also have clearly distinct chemical formulas - the phospholipid membrane of the cell, or DNA itself for instance (C232 N92O139P22). So, there is a clear pattern. Should we deny that it exists? 3. even chemistry, in essence, is based on observing a phenomenon, except at tiny scales.
I also get the comment that from my stance would follow 'life is a force present in the living'. This results only if you shift from my definitions to those of others, since I have clearly defined above biological organisms to be distinct from 'living'.
The only reason I made this distinction between 'life' and 'biology' in the first place, is because there seemed to be confusion about 'life' as a physical force vs 'life' as a common denominator for the systems with cells and DNA. Therefore I divided the two into life - a physical force present in biological organisms, but that does not exist and thus is defunct- and biology - organisms with cells and DNA.

Whether or not we can explain everything with the laws of physics and chemistry, and not with the verbal mess and uncertainties that biology introduces, the term biology has a reference to a well-defined category, at least in my definition. But, by all means, if the superior method of physics and chemistry can fully replace the METHOD of biology, then it should so. So, biology could be considered defunct, not as a term, but in its methodology - if chemistry, physics, and mathematics can prove their worth in predicting biology. In theory it is extremely likely, but in practice, the biological method still prevails.
As a conclusion, we say that 'life' - as a physical force present in biological organisms but not in non-biological things - is defunct. The term 'biology' - organisms that have one or more cell and have genes coding for their proteins - is not defunct. But the 'biological method' - verbal and not always

consistent - hopefully, one day, will be defunct, for chemistry and physics can give more precise predictions.

To say that the biological method is defunct is very exaggerated of course, but it corresponds to what Libb's vision is - and, I believe, also is one of the main reasons he even adheres to the vision that biology is defunct: because its method can, in theory, be replaced by the chemical method, and of course, because precise quantitative statements are always to be preferred over vague verbal statements. Yet this is only in theory; if it is to succeed, a very long path must be traversed, and it might not be that useful to actually traverse this path, regarding the history of success of the biological method in neuroscience and medicine. Very high success has been shown in biology lately – since for example, we have almost entirely reverse engineered the brain, mapped the entire genome, have stem cell research, can clone, etc. etc. So perhaps the biological method is just fine. The chemical method works fine when dealing with the scale of some simple molecules. But at increasing scales, it becomes somewhat impractical to use the chemical method: it would be like applying string theory to chemistry (possible, but not a very efficient way of getting valuable information). Even if it would appear that the bio-*method* would one day be defunct, the *concept* biology still refers to a category with clear characteristics (the two criteria of course).

Libb: *The following page has some 1936/2002 comparison images to help you to see that your "cell-centric view" of things is the old 1936 method:*

http://www.eoht.info/page/CHNOPS

Correctly, you need to catch up the new 2002 model.

David: *this illustrates that biology can be explained in terms of chemistry. so the biological method may be defunct - or still under slow degradation - but the term biology itself refers to a well-defined class, you know the two criteria.*

What this story tells us is how important definitions really are, and how important it is, for the sake of conversation, to define them clearly, and to apply them correctly. Libb by now still cannot apply the definition. Libb, after about 20 times having been told the definition of biology and the definition of life , still cannot see that biology is *not* 1. An undefinable category – as he many times proposes. 2. The same as a physical law that is present in organisms consisting of or having cells and having DNA, but not in others – this separate force we would call life force or a soul 3. falsified of being existent by saying biology is determined by chemistry- this would imply no categories except for the categories in the periodic table can exist, thereby making any conversation impossible. Try to refer to every word I wrote on this page by chemical formulas, and see how reasonable that stance is. Libb then should update his site, and replace all words referring to things as chemical formulas. Or, even worse, since atoms contain even more elementary particles, we should describe everything in quantum-mechanical ways… Or even strings… etc.

Well, at least the initial definitions were right this time, the only problem was that after three lines he decided to completely shift definitions by defining biology as the study of life (not what any of my definitions claimed) and subsequently erroneously claiming that from my view would follow that 'life is a force that moves life'. First, let us examine why Libb cannot use definitions. An explanation why Libb might refuse to see what I mean, is thinking in or-or-terms, instead of and-and. Both 'animate science' and 'biology' are well-defined concepts, that in no way refute each other's definition. Another explanation is that Libb might really just be blinded as to what I am saying because he has been spending much time believing biology is 'the study of life'. Although that is the definition of *biology*, both Libb and Tesla had, implicitly, defined *life* as something it was not to biologists, defining it as the force that is present in the atoms of organisms and not in those of things, so that everyone *had* to agree: life indeed is a religious term – there is no such force where physicists and chemists are looking for that is exclusive to biology (although there have been some proposals in terms of physical laws). Libb consequently referred to biology as something that it was not: the study of forces that are present in the atoms of organisms, but not in things – in short, biology was now referred to by to by him as "the study of things with a soul, or a 'life force' ". Hence I wanted to make clear what Libb was actually debunking, namely this physical 'life force' where physicists have been looking for, but not biology – since biology is *not* 'the study of things that have a soul or a life force', check with any scientific definition of biology!

Libb often refers to the 'fact' that every definition of life has been fallacious. My response: Definitions are agreements. they cannot be fallacious. relating to the life debate, if you see an instance where your notion of life does not corresponds to the definition given, then that is 1. merely a sign that you have your own definition - i.e. a sign that you do not agree on how the definition should be. not however, that the definition is fallacious. 2. a sign that you should restructure the examples which belong to the category implied by the definition. for instance, if you consider erythrocyte as a sign that a definition of life -having cells and DNA- is faulty, then this is a sign that 1. you already imply a definition of what life is 2. what you should do in this example is, if you decide to apply this definition, cut out the erythrocyte as not being alive.
in mathematics for instance, we can define 2 as the number that is neutral to multiplication, and 1 as the number that doubles a number after multiplication. while this definition is not according to agreement, we can perfectly work with this definition. so if you then say that this defintion is 'fallacious', that is incorrect. you then just still use your definition (1 as neutral number for multiplication and 2 as number for doubling after multiplication). the only things that matter are 1. stating what your definitions are, when you are explaining some theory 2. applying precisely that definition and no other. if you then argue that we cannot apply the definition of life precisely, because there exists no such precise definition - then that is merely because you have a fixed series of examples of living beings (as defined by earlier definitions) in your head.

Yes, definitions are very random. Why don't we use '!' for multiplication? Why don't we use X for the term life? Indeed, the fact that Libb considers many definitions fallacious merely refers to the fact that he uses another definition as a reference. For instance, if we define 'life=intelligent', then Libb would say 'a computer is intelligent'. This reflects the fact that he uses the examples of what is usually considered to be living – this is *not* applying the definition.

Libb's definition of life- borrowed from Tesla, added with ancient notions

Perhaps the main effect that Libbs theory of life as defunct had is that the definition of life has changed – since mainly the objection of Libb was that we should show there is some force in atoms that pervades the biological and not the non-biological. My reaction to this was the only one possible: clearly defining what he believed he had debunked and equating that with 'life'. So we would have to separate from what is *very commonly* referred to – since to any reasonable person it is clear that there exists no such life force - as living organisms, which from now shall be called 'biological' organisms and *never* 'living' – which signifies from now on this force that should be present in biological organisms, but not in non-biological things – but of course in reality does not exist. Another reason why it deemed necessary to separate the two terms, was that Libb pointed that I should prove atoms are alive. Atoms should only necessarily be alive if life is some physical force, and clearly not if life would mean 'having cells and genes' – again, should I prove that my toe contains my foot? Thus clearly, a separation was necessary where the physical force now was described by 'life' (or 'life force'), and where 'biology' now as the study of 'systems that have genes that code for their proteins and that have cells'. Not only Libb has interpreted life as a physical force, but also perhaps the search of many physicists for the life force made many – me as well – believe that there is no force that is present in biological organisms that is not present in non-biological organisms. One famous quote of Tesla, who as well referred to life as a physical force is the following :

"There is no thing endowed with life—from man, who is enslaving the elements, to the nimblest creature—in all this world that does not sway in its turn. Whenever action is born from force, though it be infinitesimal, the cosmic balance is upset and the universal motion results." –Nikola Tesla

Clearly Tesla refutes the concept of life, but therein stresses the concept of determinism, saying that forces move all, that we are all (things and biological organisms) determined by these external physical forces. The concept of determinism is only relevant to debunk the concept of life, if life is to be seen as a physical force. The word 'endowed' also somewhat reminds of the in cartoons frequently posed vision that your soul goes out of the body at death. Or at least, the word 'endowed' refers to a property that should be universally present – just as well in atoms. For such reasons, it seemed necessary to distinguish between this concept and the concept usually referred to – the way *biologists* (not physicists) would define life. An interesting note to make is that in Tesla's time, the structure of DNA was not known.

One does not necessarily need to make this distinction between life and biology however, since few people actually use life in the way that Libb does. Libb tends to stress one *aspect* (or connotation) of 'life' – namely its subjective meaning to some religious and uninformed people, to see life as some 'vis vitalis'. It should be clear that *no biologist* sees his object of study, 'life', as the study of things that have a soul. So the distinction needs to be made *only when* life is to be interpreted as a physical force. We can also drop the word biology, but that would be slightly ridiculous, if its only purpose is to justify an under biologists non-existing definition of life or biology. Does any biologist define 'biology' as 'the study of things that have a soul'? No, normally they would tell you something about genes, cells, organs, etc.

There are roughly three options in the debate on definition **A)** recognize that 'free will','soul' or 'separate physical force for biological organisms' is just one connotation of 'life', and a non-utilized definition by biologists. And thus consequently both biology and life (in this case the object of study in biology) are *not* defunct. Or **B)** Assume the definition of life that Thims, Tesla and many other searching physicists implied – by means of emphasizing determinism and non-existence of the soul - and consequently conclude that biology does not define itself as 'the study of things that have a soul'. In this case, clearly, life and biology need to be separated concepts **C)** Assume the definition of Thims and Tesla and consequently drop the term biology as well– and according to Libb in this case we should use the term 'animate science'. This of course is fine in its own right, but then the category 'animate science' does not correspond to the category of biology: *everything* is moved by external forces, so everything is animate. Thus 'animate science' is in no way a replacement of the term 'biology'- a term often handy to use. In the Hmolpedia the term animated matter – the study object of animate science -is defined as 'matter (fermions) that moves or is animated, generally by the action of force (bosons, which are presumed to cause the 4 fundamental forces of nature)'. Again from Hmolpedia, we read: 'the animate system receives external forcing. When this forcing is removed, the system will degrade in accordance with the second law to a state of equilibrium. An example of a large, animate system is the earth, which can be idealized as a type of large photon mill, where the external forcing comes from, to a large extent, the sun.' So, since when is the earth somewhat corresponding to biology? When presenting such a question, Libb might say: 'that's the point, biology, a category comprising a large amount of animated objects, is not distinguishable from other animated objects.' However, it is clear that there *is* a difference, since there is a large consensus when saying the word 'biology': when going over this category, you will quickly come to the conclusion that biological organisms *are* distinguishable from 'other animated objects' by the criteria of genes and cells.

I opt for B, since it gives the philosophy of Tesla and Thims their place and correctness, but at the same time stays with the correct definition of biology. The question then remains what the definition of death is. Well, if we apply the definition of life and take death as its antonym, then we arrive at the following definition: 'when the life force, or soul, or other physical force that is only present in biological organisms, leaves the body, supposedly going to another host body (reincarnation), or to the afterlife'. Clearly, death in this definition is a religious, defunct term: of course in this definition death is a term that bears no relevance to scientific understanding. With regard to the antonym of biology or biological organism we have to specify the process of degrading, e.g. 'biological degrading', meaning a decomposition of the body, where the cells degrade into smaller pieces that composed the cells. These smaller pieces, for not containing full DNA, nor containing all the necessary characteristics of the cell, are now non-biological molecules.

Three more comments need attention.

Firstly, if we adhere to the philosophy that scientific studies concerning a field which could as well be explained by chemistry, then we would have to declare all those fields as defunct: mineralogy, geology, neuroscience, biology, psychology, sociology, etc. Also, chemistry might analogously be explained by quantum physics, or even string theory, which would mean that string theory would be the only valid approach to do science. If you don't think that's a little unpractical – why else would we change terms and definitions if those terms and definitions weren't also handy in use, and, well, distinctive ? – that does not mean that it is in any way preferable to throw all the terms and methods for other scientific disciplines in the garbage can. I don't think it is necessary to approach everything via chemistry, as in many cases it might not be the most efficient way of obtaining information (for instance, talking about a heart attack is pretty useless to do via chemical equations, since it would be quite unnecessarily complex and it would contain much redundant information). I am strongly of the philosophy that different contexts and different functions require different approaches: whereas the human molecule theorist (correctly) assumes that a human can be described as a relatively stable molecule and derives predictions from this, it should be clear that many other approaches have success in predictions. On the one hand we can define us as a molecule, on the other hand, in other contexts this knowledge is inapplicable and is the worse method of the two. For instance, try to explain varicose veins via the human molecule approach. Ok, what about the cellular molecule approach? Well, none of the evidence has shown that any of those molecule methods are better in explaining varicose veins than the biological approach (relating it to factors as blood pressure, arterial dilation, etc. instead of referring to elementary composition). This to make clear that different approaches should be used for different problems, although different approaches can and should be used for the same problems as well. Afterwards, we can decide which method has consistently shown to be of higher predictive power, and we can favor one approach above the other on some specific problem – yet still not use this approach solely, because methods can be improved, and coincidences can occur. One cannot say we should not use the term biology, just because in theory, everything should be chemically determined. Evidence that biology is alive and well is the medical science, which looks at larger wholes such as organs and looks at its functions.

Secondly, by defining life as a physical force present in biology, and biology as the (study of) organisms that have cells and have genes coding for their proteins, I have actually given in to a somewhat weird assumption of Libb: Libb refers to life as equal to the soul or a physical 'life force'. In all of his writings, Libb implies that we should prove that atoms of biological organisms are alive , and also that 'life is a religious term'. This clearly results from not defining what biologists think of life, but rather it reflects that Libb has been studying all failed attempts of physicists to explain life by a yet to be discovered physical force. So, by defining 'life' as this force physicists were looking for but will never find, and 'biology' as the organisms with cells and genes that code for proteins, then I have shown what it is that is 'defunct' (the physical force falsely assumed to underlie biology), and what it is that is 'not defunct' (the category 'biology', (the study of) all the organisms that have genes coding for its proteins, and have cells). If Libb on the other hand defines 'life' as he does, and 'biology' as the study of life, then indeed biology is a defunct term – but this results only from a highly unusual definition of life (a 'life force', a soul), and thus in this case, biology. In the definitions that Libb uses - 'life' as in the definition we have agreed upon, and 'biology' as the study of life- then indeed biology is a study of a defunct term. However, 'the study of life' is clearly *not* my definition of biology.

Thirdly, Libb has mentioned the "CHNOPS-model of 'animate organisms'". He has made the following synopsis at the Hmolpedia on the subject of animate organisms:

> In science, an **animate organism**, as contrasted with an inanimate organism, is an organized body with property of chemical animation, and more fully is animate atomic geometry (animate molecule), type of animate matter, or animate chemical species, such as a walking molecule, above that in movement complexity than the simple mechanical automaton, such as windmill, generally tending to be moved by chemical processes, e.g. exchange force based induced movement or free energy driving force methods of system transformation, rather than simple mechanical or gravitation animation processes, such as is the case in the hydrostatic pressure induced movement of the Hero automaton.
>
> The 1620s <u>reaction automaton theory</u> discussions of French philosopher Rene Descartes (IQ=195) along with the 1950s <u>free energy electochemical automaton theory</u> of John Neumann (IQ=185) may well facilitate the differentiation of the various types of "animation", e.g. Greek automaton animation, AI-based animation, etc., as compared to the animation found in a moving reactive human.

Examples

> The simple 3-element retinal molecule is a simple example of a animate organism, one moved by the exchange force of the photon; the 15-element bacteria molecule is an intermediate in complexity example of an animate organism, one that generally requires some type of agar

medium to function, DTA, driven in surface-attached motion by a heat, is another example; the 26-element human molecule (person) is a more complex example of an animate organism. Other examples include: fish molecule, cell-as-molecule, the two-legged kinesin protein molecule, that walks along microtubules while carrying cargo.

Discussion
The term "animate organism" is an upgrade from the defunct term "living organism", in that a body can possess animation or reactivity, but not life—in the sense that "life" is something that does not exist, in the famous 1925 words of Nikola Tesla.

To explain another way, in the hmolscience periodic table perspective, a zirconium-based organism, e.g., will not have the property of "animation"; such chemical species may exist, and may have the property of trajectory or unidirectional movement, possibly even spin, but not complex animation. Conversely, carbon-based organisms, as embodied in the out-dated term "carbon-based life", will tend to possess the property of animation; but not all carbon-based entities, graphite being one of many examples. Those carbon based animate organisms in possession of the so-called CHNOPS-based or "CHNOPS system" set of core elements, while in the bound state of animate existence, may well then be classified as types of "animate organisms" and as such function to proactively take the place of the now-defunct terms of olden: living organism or living system, which have no hard physical science basis, but only mythological basis; hence resulting to do away with the "unbridgeable gap" model of the last century.

Other
German mathematical philosopher Edmund Husserl (1859-1938), according to Google Books search results, seems to have been a dominate advocate of the term "animate organism". [1] Danish natural science philosopher Claus Emmeche uses the term "animate organisms" within the loose context of thermodynamics discussions in his 2004 article "A-life, Organism and Body: the Semiotics of Emergent Levels"; inclusive of all the other "life terms". [2]

Whereas the biologist emphasizes that biological organisms have unique properties in terms genes and cells that can be seen as functional elements, the thermodynamicist and chemist will emphasize concepts such as free energy, the six elements CHNOPS. We cannot say either of them is wrong, or that one or the other is defunct – both of them are correct, but on different levels, because indeed both concepts are (or at least *can be*) well defined, but different not only in their approach, but also in their scope: biological organisms are a subset of the animate molecules which at the macro-level have properties such as genes and cells. Biology is only defunct if indeed you equate biology with the study of 'organisms that have a soul' or 'organisms that have life' if you view life as a physical force. In scientific terminology, biology does not equate with this, but instead it equates with organisms that have cells and genes. According to the Wikipedia definition, biology has 5 fundamental axioms:

1. Cells are the basic unit of life
2. New species and inherited traits are the product of evolution
3. Genes are the basic unit of heredity
4. An organism regulates its internal environment to maintain a stable and constant condition
5. Living organisms consume and transform energy.

This clearly has nothing to do with 'soul' or any physical 'life force'. Some may view these terms as a connotation of life. Biology is more of a scientific term and, in my opinion at least, has less connotations. Life is employed in many contexts, such as 'social life', 'I can live with it', 'get a life'.

The Hmolpedia gives us more insight into CHNOPS:

In <u>animate science</u>, **CHNOPS** *is an acronym short for the six dominant elements common to animated organisms, namely: carbon, hydrogen, nitrogen, oxygen, phosphorus, and sulfur; the other dozen to twenty or so elements common to animate organisms sometimes referred to as trace elements, although the distinction is far from exact.*

Old system

It remains to be determined who exactly introduced the so-called "CHNOPS" model, but it does seem to have something to do with the <u>Hill order</u> scheme for arranging atoms in a molecule developed by Edwin Hill in 1900, according to which:

1. For carbon-containing compounds, carbon (C) appears first.
2. Carbon is followed immediately by hydrogen (H), if present.

3. Compounds are listed by increasing number of atoms.

4. All non-carbon element symbols follow in alphabetical order, and within alphabetical order are listed by increasing atom count.

In this scheme of things, in alphabetical order, CH would be followed by N, O, P, S, and so on, hence the acronym: "CHNOPS".

The term CHNOPS dates to at least 1936, wherein one article in particular defined "Chnops: six chemical elements are essential parts of <u>protoplasm</u>, the <u>living substance</u> itself. These are carbon, hydrogen, nitrogen, oxygen, phosphorus, and sulfur." [1]

The term "CHNOPS system" was introduced in a 1964 US National Bureau of Standards report entitled "Preliminary Report on Survey of Thermodynamic Properties of the Compounds of the Elements of CHNOPS" by a group of researchers led George Armstrong. [6] The following is an example citation from Henry Swan's 1974 Thermoregulation and Bioenergetics: [4]

*"But a biochemistry could emerge in which life is powered by the * This small group of low molecular-weight 'core elements of life' has been dubbed the '**CHNOPS System**' by Armstrong, et al. (1964)."*

In 2005, American electrochemical engineer Libb Thims, following Hill order protocol, posted the following online first-draft listing of the molecular formula for one human: [5]

Note 6: in accordance with Hill Order, we arrange the (first approximation) formula for the human molecule as below:

$$C_{E28}H_{E28}N_{E27}O_{E27}P_{E25}S_{E25}Ca_{E25}K_{E24}Cl_{E24}Na_{E24}Mg_{E24}Se_{E24}Fe_{E23}Co_{E23}$$
$$Cu_{E23}F_{E23}I_{E23}Zn_{E22}Si_{E22}Mn_{E20}B_{E20}Cr_{E20}V_{E20}Sn_{E19}Mo_{E18}Ni_{E16}$$

Further refined calculations, however, began to give way to the view that the old Hill order system (alphabetical) was an an inconsistent way of ordering atoms, particularly when dealing with a 26 atoms in one molecular formula. The new system, outlined below, lists atoms via decreasing atomic count in the molecule.

New System
(2002)

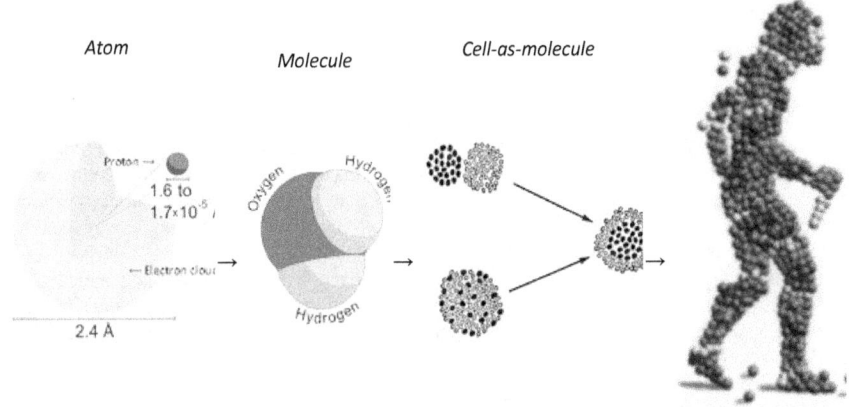

The new system model is depicted well by the 1993 "cell-as-molecule" approach, pioneered by English physical chemist Lionel Harrison, and the 2002 "human-as-molecule" (human molecule) approach, pioneered by American limnologists Robert Sterner and James Elser, according to which organisms are defined as individual abstract molecules, each defined by a characteristic molecular formula: 22-elements according to the <u>Sterner-Elser human molecular formula</u> (2002):

$H_{375,000,000} O_{132,000,000} C_{85,700,000} N_{6,430,000} Ca_{1,500,000} P_{1,020,000} S_{206,000} Na_{183,000} K_{177,000}$
$Cl_{127,000} Mg_{40,000} Si_{38,600} Fe_{2,680} Zn_{2,110} Cu_{76} I_{14} Mn_{13} F_{13} Cr_7 Se_4 Mo_3 Co_1$

or 26-elements according to the <u>Thims human molecular formula</u> (2007); a view according to which the old 6-element CHNOPS model becomes obsolete, archaic in a sense.

New system
In contrast to what might be called the "old system", i.e. the CHNOPS system model (1936), the "new system" is centered around animate molecular formula point of view, i.e animate molecule perspective. In the human-centric case, the human molecular formula perspective is followed, according to which there are 26 functional elements of relevance rather than 6 as described by the old system model. In this sense the CHNOPS ordering model becomes somewhat inoperable. The following, then, in modern terms, is the 26-element formula of the molecular formula for a typical 70kg (154lb) person: [2]

Different levels of definition: effective, scalar, Gestalt ; A methodological exploration

So, this was somewhat of an introduction to the concept of CHNOPS. I don't see how any of this refutes the concept of biology: one does not have to think in terms of or-or, but rather in terms of and-and. It should be clear that we are talking about *different levels of definition*. We cannot talk about the function of DNA if we use either of the two stances (CHNOPS or animate molecule). However, the CHNOPS system or the new 'animate molecule' perspective will, both in their own ways, prove us to be useful perspectives for thermodynamicists and chemists. None of these terms should be defunct, since they are all truths, albeit defined on different levels.

Try to describe why the person is no longer 'alive' (as in biologically degrading), without the use of functional descriptions. For instance, say we use only chemical formulas: we arrive at the conclusion that a description of heart failure in terms of 'the hart fails to provide the cells with blood' is the best way to approach the problem. The term 'animate science' does appeal to discard religious notions such as a soul, free will, a 'life force'- but that in no way means that biology is a defunct term, since biology does not study 'life forces, souls or free will' – and neither does it *assume* one of those. On the other hand, a short note has to be made that 'animus' means 'life, soul, breath', so it too has connotations with both life and soul. But of course, in any event, we can make agreements on what terms to use in what context – that is what categorization is all about, and that is where this conversation is all about.

Do we really have to use my distinction between 'life' and 'biology'? It depends: if one uses 'life' to signify the 'physical life force' or the 'soul', then we should clearly make a distinction between 'life' and 'biology', for *nobody* uses the term biology as 'the study of things that have a soul', but rather, biology is, for most biologists, implicitly defined as the study of 'organisms that have cells and genes coding for their proteins' – albeit that some biologists would have to think long before arriving at this solid definition. If, on the other hand, we *do* want to specify life as the object of study in biology, then we should rightfully define 'life' as 'organisms that have cells and genes coding for their proteins' and biology as 'the study of life'.

In any event, this story illustrates how important it is to specify definitions *before* we start a conversation, or before we make any conclusions on the defunct status of a term. Whether or not all things are determined by a fundamental process like free energy, there are a multitude of terms which rely on common characteristics that are nevertheless clearly distinguishable as one common category, be it by means of special molecules present, phenomenological appearance, function, movement, or etc. We can still talk about things that are inherently simply a composition of elements as being different, because of their unique composition, some unique combinations, some unique phenomenology, some unique function – this vision does in no way damage the scientific status of the visions by human thermodynamic analysis, as the two are simply descriptions (and hopefully, predictions) on different levels of analysis. Which of the two approaches will be more successful in the future, it is hard to tell, but approaches via the verbal explanations of the biology of cells, organs, tissues, the brain, etc. have been very successful in medical science. Why verbal explanations work is that you take larger molecular structures, for instance bile or urine, and you describe its effect in terms of reactions with its environment. Since this substance has that effect on that substance because if not X, we infer that its function is to X. In other words, taking the chemical reasoning into account, we can abstract why our body seems to function this way. If this chemical reaction did not happen, then digestion of lipids would be difficult, so the conclusion is that bile's function is to digestion of lipids. Via the knowledge of these processes - sometimes based on chemical equations, other times taken a more phenomenological approach – we can infer what the function is. On the basis of this function we can make classifications on the basis of this abstract property, since categorization, just as with the naming of chemical elements, rests on the fact that we make agreements on some common properties that are in some way (function, phenomenology, appearance, physical substrate, reaction) distinct from the properties of other things.

With a ball for instance, we base ourselves on the property that its phenomenology is common. But even shape one can also easily categorize in mathematical terms, making it also a sound scientific description: balls are 'those objects whose volumes are $4/3 \; Pi \; R^3$, where R is the radius of any cross-section of this object." With neutrons we classify in terms of its quark composition. With chemical elements, we base our categorization on the number of neutrons or protons. With DNA, or cells as well, we base our categorization on a unique molecular composition. So it is with biological organisms: this group distinguishes itself from the group of the things that have no genes (which 'function' is to code for proteins).

The label function is of course merely an effect, that we post hoc say to be the 'reason' why something functions this way. For instance, if the neutron did not exist, then the repulsive force between protons would cause the nucleus to tear apart. We thus infer the abstract principle that 'the function of neutrons is to cause stability of the nucleus', since our world would not exist in the way it does now. Of course, goal directedness does not truly exist, so the label function is perhaps too anthropocentric. A better label is **effect**. So the word 'motor' has an *effective definition,* since it is defined mainly in terms of its effect on its surrounding structure. This will often be named functional or practical, but this is to vague, since the words function and praxis imply purpose. Man may infer terms such as 'use', 'praxis', etc., but the nature of use and praxis is highly dependent on what, if any, goals the system has. If the system cannot conceive of the concept goal, then we should better avoid such terms. So, for more general cases, we use the term *effective*.

While I am of the opinion that biology has had a lot of success in medical science, in my publication in the Journal of Human Thermodynamics, I note some difficulties in discussing evolution and also mate selection in a verbal biological way, and I note another range of paradigms. The main idea is that evolutionary discussions on mate selection recursively assume that a certain partner was more desirable for being more dominant, a better provider or so. More important and more predictive it seems to be that both partners are chemically and genetically compatible. Also, I of course would *like* to have all sciences predict things with absolute certainty mathematics-style. Unfortunately, mathematics is sometimes too difficult to implement sensibly into most biological or psychological predictions – at least for now.

As might be apparent in my writings, I believe in the power of converging methods – many methods can attack the same problem via different angles. If the solution is unanimous we have a clear catch. Also, I believe in a multi-level approach, which implies that we tackle a problem from many levels, but we prefer the levels that are most consistently efficient in finding a solution to a problem. Perhaps this idea relates to the ideas of yet another Hmolpedian, G. Gladyshev, on the subject of 'hierarchical thermodynamics', approaching the debate via hierarchical analysis.

The method of physics > the method of chemistry > the method of biology in terms of maximum predictive precision- but this maximum is not always attained, because we might have difficulty figuring out how to quantify the terms. On the level of medicine and neuroscience, the biological method applies chemistry to identify larger structures such as organs or cells in terms of their functions (supplying blood to cells or removing wastes) and their interactions on a more apparent level (viscosity, compressibility, thickness, pressure, pH level). It in many instances does not make sense to apply the human molecule vision because other methods have shown to be more efficient (yes, medicine has been very successful without the human molecule vision), but that does not mean the human molecule, or the cell as a molecule vision should be abandoned – since these will yield predictions at other levels. The most efficient method, in a particular level and in a particular domain, should be the predominant one – but one has to leave room for innovation and improvements for the less efficient methods.

While being extremely exact in many of its predictions, most of medicine is entirely verbal in its description – not mathematical, aside from the statistical procedures to validate certain findings. In medicine, it is more important to know which macrostructures are where, and what they do in terms of function and reactions, then it is to somehow quantify what they are doing. For instance, it is more important to describe that a virus has entered the body. This fact alone – macromolecular composition to *identify* that it *is* a virus and proven *effects* on the body (by repeated empirical evidence; namely illness) – is more important than a quantitative description of any kind. Another instance is cognitive factors. Mathematics cannot yet describe what happens in the process of 'knowing'. By this process of knowing I mean, when someone says 'the queen died today', you now have acquired this piece of knowledge. The sum of all these pieces is the whole 'mind file', which determines all what we think. What we think in turn, determines our actions and our choices. Mathematics and physics, I believe can explain this one day (by showing what memory imprinting is), but that still does not mean that I should refer to knowing this thing via this mathematical explanation, for instance when I want to talk quick and everyone should know what I am talking about. This to say that different contexts require different terms, methodologies. Using the quick way of, say, 'knowing', it becomes easier to build further on existing knowledge. We can describe a process like memory imprinting mathematically, but we cannot contain what reactions those things elicit unless we know the entire mind-file and its connections to neurotransmitters. So making mathematical predictions is hard and complex in this case, and we might want to simplify 'knowing' or 'memory imprinting' into 'the memory that his father molested him', and see this as an elicitor of a somewhat painful feeling. At some point, thus, the mathematical approach becomes too difficult – and we might want to take a shortcut. Although less efficient if we really had a working mathematical model for such complex matters, it is clear that we don't have such a working mathematical model – for the process of memory imprinting, even in one instance, is extremely complex. So in essence, another approach is more efficient in practice: because of the inherent complexity of a mathematically sound description, mathematical description cannot come further than, say storing one memory. Mathematics can thus not explain why this person made this choice - but only because human minds are too limited. So the future of course holds promise of many mathematically sound predictions of human choices. Should we then at this moment throw away cognitive approaches? I think not, since neuroscience, AI, and cognitive psychology go hand in hand, and they have shown much success (see the progress that has been made in reverse engineering the human brain, and the progress in intelligent machines), as does biology and medicine (see the many cures for diseases, transplants, cloning, etc.). So, when do we throw away these approaches? Only when the results show that mathematics and physics can solve these problems much better than is currently the case.

It does seem desirable to give definitive quantitative predictions. So definitely try to progress as much as possible in mathematics and physics, which are of course the very fundaments of the scientific enterprise. In this way, we must maintain *ideals* that mathematics and physics, or even cellular automata can replace the biological, verbal method – which sometimes uses very vague definitions, especially with regard to behavior. I talk about this more in the next chapter, wherein I discuss several methods for arriving at truth, and where I talk about the downsides of the biological method with regard to mate selection predictions.

A digression on the question of goal-directedness

In any case, the question if 'goal-directed' really exists is again very dependent on the definition. One can make analogies that go somewhat further: what if the lamp and the television are to us, what we, or the whole world, is to some other highly intelligent entity? What if we were, like we do research on animals, in a big simulation to see whether this or that environment factor has this or that effect? In this way, lamps or televisions could be seen as product of a goal-directed creation, since it was the creator's intent.

In another sense, even taking to account the possibility of some intelligent creator, one could argue that this creator is just as well determined by the laws of physics of the world around him that determine his behavior and his intent, so that one can speak that *nothing is 'goal-directed'* in the sense that the laws of physics dictate without some other, perhaps immaterial power that would be called '(Free) Will'. So when discussing the question of 'goal-directedness', one must ask: 'who is the agent pursuing a goal?' Is our world in some sense related to a goal in the perspective of this hypothetical intelligent being? Yes, since this hypothetical being uses us to gain some advantage for himself. However, this hypothetical intelligent being will always reside in some other distant place, which has physical laws that determine his behavior. If one was to take *all* that existed, then we would come to the conclusion that existence (as in 'the sum of all that exists') is not a goal, since it has no external agent to which this existence can be related to any goal. If one starts to postulate some being beyond 'existence' as here defined, then one would come to the conclusion that this being does not exist; hence we have a reduction ad absurdum. So in essence, 'existence' is by itself not goal-directed. To some agents however, parts of it can be goal-directed – meaning that they have some function in the pregnant sense of the word ('serving some agent a particular goal to satisfy a particular need'). In the chapter 'our origins', more of the debate on creation and existence.

A conclusion on the 'life' debate

I want to quote another post by Jeff Tuhtan and consequently by Libb, in a somewhat older thread in the Hmolpedia:

Jeff: *I think the fundamental question sadly comes down to a matter of semantics. Why don't we get rid of the terms 'living', 'biotic', etc.? Well, probably because of their high functionality in our everyday 'life.' If we could come to an agreement that when scientists refer to 'living' organisms, they have a certain spectrum of chemical complexity in mind, then I could accept that for the meantime. It will take generations before the depth of the concept trickles down into society as an acceptable level.*

This is *one* of my stances on the question: practicality. Terms such as life and biology have much, and variable uses and therefore in many ways it holds meaning, albeit in different ways dependent on the context.

Libb: *I agree with you that in large part the issue is a matter of semantics, but at a deeper level the true "fundamental question" is a matter of morality, purpose, theoretical ideas about "afterlife" (or how the day-to-day choices made by a person reverberate in time, in society as well as in the dynamical movement structure of the universe), when he or she is gone (ceases to exist as bound state atomic entity), as well as cherished ideas concerning "free will" and "freedom".*

Another stance of mine is 'non-existence of afterlife, soul, immateriality, free will' - corresponding to the answer to the fundamental question that Libb would as well adhere to. The latter thing between quotes (freedom and free will) we would/could/should refer to as life from now on – in this way, 'life is a defunct term' and 'there is no thing endowed with life'. Related to this stance of course is the fact of determinism – a philosophical stance I find very reasonable, given the fact that all can be explained and, in theory, predicted by physical forces.

Another point I take is to separate biology from the term life. Although commonly defined as 'the study of life', it should be clear that biology, in no scientific definition, is defined as 'the study of things that have a soul/free will/ a physical life force'. Thus biology should be best referred to as (the study of) 'organisms that have genes (DNA and/or RNA) coding for their proteins, and that have one or more cells', or other related definitions. There is another option, however, and that is to simply separate 'life' from 'life force'. In this way, life would indeed refer to the object of study of biology, whereas 'life force' would be the defunct concept.

Chemical determinism does not seem to me any reason to abandon the concept of biology, given the spectacular success - by looking at some visible large-scale properties, the function or effects, macro-composition, etc. – in the medical sciences. For instance, organs are usually described in terms of their appearance and their function – for instance digesting lipids. Or for instance, say you want to approach psychology chemically or mathematically. Given the complexity involved in the mathematical description, we will need a googolplex amount of years to solve such a problem. However, with psychology, you can for instance use the historical-verbal method: if I say 'bitch', she will become mad. That is quite a simple solution, yet overlooked by chemists, who think it more efficient to describe this in mile long chemical equations. I do not deny that it is a very rigorous approach, but nevertheless they take way too long for us to consistently rely on these methods for this purpose. Interestingly though, these truths will be different than the truths derived from the historical-verbal method, since they are described on a different level – and perhaps these truths are more 'elemental'. One can however not deny the truth of a sentence as 'she will become mad', or, with regard to political speculations, it seems more useful to say for instance 'if we attack, they will attack' and not look for the chemical substrate of this process of conditioning – to me one of the fundamental rules in psychology, which also happens to be in line with a standpoint of determinism. In short, different methods serve different purposes.

It is *not* my goal to change the definition of either life or biology. The main reason I even make this weird distinction is a response to Thims, who saw life as a physical force that is (by religion, or non-scientific philosophies) assumed to be in the atoms of biological organisms and not in those of things. This physical force could also be referred to as a 'life force', a 'soul', or 'free will'. In this sense, life is a 'defunct' concept. It was the same 'life force' Tesla envisioned when saying the famous quote, which stressed determinism and the absence of a physical force separating 'life' from 'non-life'. However, if consequently, one would still define biology as the study of living organisms, then one would come to the conclusion that biology doesn't study anything, since any biologist knows there is no such separate physical force, or at least, no scientific definition of biology uses the concept 'soul' or 'life force'. Nowhere have I encountered, in a definition of biology, the fact that these organisms should have a soul, or a separate life force. Most definitions would refer to genes as a hereditary component, cells as fundamental building blocks, the existence of organs or organelles which perform specific functions to an internal system. Thus, it seems necessary to separate the concept of life – the physical force that physicists have been looking for - and biology, the study object of the biologists, pretty clearly nowhere related to the soul. So what has happened mainly here, is not that I have changed the definition of biology, but mainly that Libb has changed the definition of life, or at least accentuated one of the religious sides of this concept. To most biologists however, life would not entail a 'soul', or 'atoms that are alive', or etc. So in this sense, life can only be made defunct by emphasizing one side of the definition that holds no meaning to biologists.

Terms Libb has proposed – besides referring to life as a life force that pervades the atoms - seem inadequate to replace the word biology. For instance, animate science is the study of animate systems. 'an animate system is a system on which work must be done by the surroundings, and which via interaction with its surroundings must be subject to periodical increases of its free energy such that it may perform work in changing its internal configuration, possibly in response to changes in its surroundings.' As we have seen above, even admitted by Libb, in this definition the earth is animate. Even my car, when an external force is applied to it, can perform work. So, the term animate refers to a much broader category than biology. As seen, we can differentiate between types of animate systems. The car, having no DNA, nor any cells, is to be contrasted with a human, which has these two properties plus properties as 'can reproduce', 'evolves via genetic adaptations', 'has intelligence'. The latter are not core characteristics since computer viruses reproduce and AI is clearly intelligent. The core characteristics are 'has, or consists of cells', 'has genes coding for its proteins', and perhaps also 'evolves via genetic adaptations'. Another concept Libb likes to use is 'existive'= existent + reactive - to indicate what other people call 'alive'. This word is even less distinguishing than the commonly used words: *everything* that exists is also reactive! Reactive is a gradual contrast – even rocks are not fully inert, look deep into the smallest scales and you will find atomic reactions. What other word, then , should we use for 'alive'? Well, there are some other options here.

A) we can refer to 'the force present in the atoms of biological organisms but not in non-biological things' as a 'life force' or a 'soul' instead of 'life'. While Libb and some other physicists *did* define this as 'life', it should be clear that 'life' in this sense is not the object of study in biology! So, if we choose this option, 'life' will refer to what it used to be (organisms with genes coding for its proteins and cells), and biology will be correctly termed the study of life. Death then corresponds to the opposite of life, meaning a biological degradation, until the system has vanished into other organisms or into non-biological parts.

B) We can keep the definition of Libb, so that 'life'='life force', and consequently conclude that 'alive' would be somewhat inappropriate to use for anything that is existent. 'Biologically coherent', then, should refer to what we consider alive, whereas its antonym 'biological degradation' should be used to refer to 'dying'. The term 'degradation' is used to accentuate that it is not a sudden process - which would let us think that, according to some religious beliefs, with 'death', the soul suddenly leaves the body, which is of course not based on any evidence. In case B, of course, we need to make clear that 'life' is not the object of study of 'biology' – so that biology is *not* the study of life.

A Gestalt approach to the debate of biology vs chemistry

Interestingly, not too long after the lengthy conversation on the definition of biology and life, Jeff Tuhtan, posted about the meaning of 'Gestalt' in the texts of Goethe. The word Gestalt indeed is very related to the conversation, since 'the whole is more and different than the sum of its constituents' – the premise of Gestaltpsychology – warns us we should not use chemical equations for higher level. Not so much that the rules of chemistry do not apply, but that the same constituents can form very different wholes, and that very different constituents can form very alike wholes. For instance, while the human molecule approach assumes we are just this molecule $C_{E27}H_{E27}O_{E27}N_{E26}P_{E25}S_{E24}Ca_{E25}K_{E24}Cl_{E24}Na_{E24}Mg_{E24}Fe_{E23}F_{E23}Zn_{E22}Si_{E22}Cu_{E21}Be_{E21}I_{E20}S_{nE20}Mn_{E20}Se_{E20}Cr_{E20}Ni_{E20}Mo_{E19}Co_{E19}V_{E18}$, for instance – as already posed in 'criticisms of science' - it should be easily refuted that Gestalts are different than the sum of their components by synthesizing this molecule, and consequently concluding that we have not made a human. I would be happily falsified however if this does yield a human. This difference of Gestalts made out of the same components means there are other factors than the comprising 'chemical elements'. It is more how which elements bond with each other – I do not believe us to be one molecule, but rather a whole bunch of molecules, who react to each other's presence. Adding a specific molecule to my body will not radically change what I am. DNA on the other hand, when slight deviations occur, will drastically deform my body and functions. DNA it seems, has a way of organizing protein synthesis and consequently the structure of our body. If the whole human molecular formula would be the same but the one would contain DNA and the other would not, guess what would happen? It would not be a human.

Enough on DNA, the word Gestalt is really important, since moving from bottom-up is what chemists usually do. People on the extreme bottom-up approach, claim that no thing can be categorized as being different if it is comprised of roughly the same elements. Well, lets think, what other factors are there?

Volume : how much volume does the system contain. Try to fit the human molecule in a very small system and the result will be different from the one in a very big system – due to high pressure in the case of small volume.

Temperature: humans keep a constant temperature of about 37° C. If not, we 'die'.

Structure : how exactly the building blocks (the chemical elements) are ordered within a Gestalt (a larger system, for instance a human) – corresponding to a multiple-molecule view as mine where in a system many reactions among many different molecules occur, resulting in 'functions' without which the system would degrade; but also a DNA-determined view of the human structure: DNA is responsible for synthesis of proteins which are important structural elements

These 3 factors are necessary 'Gestalt factors' to arrive at the same Gestalt, given the same elemental building blocks for the human molecule. Others exist, of which some might come into discussion.

A related Hmolpedia figure with regard to structure is William Jones:

In philosophy, **William Thomas Jones** (1910-c.1990) was an American philosopher noted for his five-volume *A History of Western Philosophy*, the first volume of which devotes considerable discussion to the formation of Greek atomic theory, e.g. Democritus, Empedocles, Epicurus, Lucretius, Heraclitus, and the so-called "atomists" in general. The following, in respect to the human molecular hypothesis, being a representative example:

"The only ontological difference between men and say, billiard balls is the degree of complexity in the groups of atoms involved. A billiard ball is being bombarded by atoms from the cabbage just as I am, but it does not perceive the cabbage. Why? Because none of its atoms happen to be grouped into that configuration of atoms we call a mind."

The configuration or structure – even if its sum of atoms were entirely the same - matters of course, and gives rise to special qualities. This is exactly what the Gestalt-approach does, assuming that a different configuration of the same elements gives rise to a very different thing. We have already made the hypothesis that synthesizing the human molecule would not lead to existence of a biological human – we thought DNA was much more important to determine structure and function, then simply a gross elemental formula. Sure, the elements in a human body are pretty consistent, but without genes responsible for protein synthesis and cellular division, we do not arrive at 'a human being'. It seems to turn out different configuration of the same elements gives entirely different structures.

Another gestalt factor is of course *form*: we can reorder a structure so that it preserves the same elements, yet that its form changes – even if its volume remains constant.

A gestalt term to replace 'death' would be 'deterioration' – the elements that once gave rise to a relatively constant gestalt, have now been separated. A gestalt term for 'birth', would be 'construction', where the elements gradually give rise to a new gestalt with distinct organs and macro-characteristics. Both these terms again emphasize the gradual nature of these transformations. But of course, in the case of birth, we may define it with a sudden feature as 'leaves the uterus'...

In biology, one would say evolution is the transfer and gradual change in genetic material. However, more broadly, one could define evolution as being an 'increase in molecular complexity of Gestalts'.

Holism vs Reductionism – another Hmolpedian debate outside the Hmolpedia, with Jeff Tuhtan

The following article was posted by Jeff, on my website – which you will love to visit! – bossensnonfiction.com.

Reducing Holism – Why the Argument for Reductionism is not Against Holism

Posted on October 11, 2012 by jtuhtan

By Jeffrey A. Tuhtan

We begin by assuming that the universe consists of a large number of finite interacting entities.

Reductionism
Often, arguments for or against reductionism are confused with determinism, which itself is just another way to embed causality within the larger framework of physics. However, it can also be stated that reductionism is truly heuristic for uncovering the fundamental. I tend to agree with Mayr that scientists should refer to this heuristic as analysis, as it removes some confusion.

It is only fair to state that as a principle, reductionism has been a success insofar as physicists continue to find entities more fundamental than their predecessors. But being more fundamental going down does not necessarily equate with fundamental gains going back up. This does not change, regardless of the number of entities considered. Where is our equivalent to a chemistry of ecosystems?

Taking my childhood goldfish as an example, we cannot argue against the fact that it is indeed made of the stuff of the universe, those larger number of finite atomic entities, and thus can be reduced to being the causal resultant of its constituents. But how small do we wish to go? There exist yet smaller and more fundamental subatomic particles and beyond contemporary experimental physical scales, a pantheon of subatomic theory. That's reductionism.

Holism
The arguments for, as well as those against holism are similar to those surrounding reductionism in that they deal with misunderstandings and misrepresentations of the concept of holism itself. Holism is not emergence, which is the concept that "the whole is more than the sum of its parts". Although catchy, this statement can only hold true if we cling to magical thinking and choose to ignore the first law, which has up to this juncture in time held up quite well as a basic tenet of contemporary physics.

Holism requires the recognition that entities and their interactions occur in a type of primal, unitary system. Anyone who has ever seen a marble, a parrot, or a pie has experienced the joy of understanding holism in its entirety for they have experienced first-hand a whole marble, a whole parrot, and a whole pie. In modern scientific terms, holism is nothing more than the affirmation that we exist in a universe. Thus holism is the acceptance of the term universe as the most fundamental expression for the entirety of our physical reality.

It will be shown shortly that "how big" and "of what" this whole consists is not really important, only that all entities which interact are part of this larger and more fundamental system. If all of the matter-energy in the universe existed in a giant parrot, it would certainly be an odd universe, but that would be the whole. That's holism.

Proof that Holism can be Reduced, and Reductionism can be made Whole.
The proof that holism and reductionism are fundamentally the same is as follows:
Imagine a complete system. The system can be finite or infinite. In our case, we define the system as

the universe at large. Now if we were to divide this system into thirds, and then put them back together again, we would find that no matter how hard we tried, we could never distinguish the difference between the sum of the thirds, 0.999 and our original, entire system. Some may think, "wait, but what if this whole is really itself just another part of some other system and not really a whole at all?" Please return to this issue, which has already been addressed above in the section on reductionism.

Now imagine a powerful science which allows us to completely reduce our universe into all of its constituent entities. And further, imagine applying this same science to put the universe back together again, considering it as the "sum of its parts". No emergence here, just Howard Hughes' bear cat on a truly grand scale. What we find is that in the end, even for universes with only three entities, there is truly no difference between the reductionist, reassembled 0.999 universe, and the holistic, singular one.

This reminds me of an old joke:

A reductionist walks into a holistic bar. He pulls out a marble from his coat pocket and bets the bartender 10 dollars that he can break and reassemble the marble as if it had never happened.

The bartender takes the bet and shakes his head in disbelief as the man smashes the marble and painstakingly without any glue, reassembles the marble, good as new.

A week later the reductionist returns with a parrot, and bets the holistic bartender 100 dollars that he can create a parrot from it base constituents.

Once again, the bartender takes the bet and watches in amazement as the man pulls out a beaker and some flasks containing various chemical mixtures, creates a fertilized egg which then grows and becomes a parrot.

After this last act the bartender was sure he would never to see the reductionist again. But about a month after his stoichiometric parrot trick he once again shows up at the bar, this time with a pie. He asks the bartender if he wants to see him reassemble the pie from its base constituents. The bartender smiles and cuts the pie into three pieces...

The reply I made to this blog was as follows:

"However, it can also be stated that reductionism is truly heuristic for uncovering the fundamental. I tend to agree with Mayr that scientists should refer to this heuristic as analysis, as it removes some confusion."
very correct, people who for instance consider a human's personality = his brain, or a human as, 'just' a clump of molecules, are considered to be 'reductionist', while in fact they are just doing an analysis. The even smaller levels (subatomic) are too difficult for now I guess, to relate them to human identity.

Personally on the "whole is more than the sum of its parts" issue: the whole is not more in terms of its constituents, but nevertheless the behavior of the whole can be analyzed using different tools, since the whole will behave different than its separate constituents. Also, it is clear to me that same constituents can build different wholes by differential arrangement. Consider a chair, which has 4 legs on its base. If you cut one leg and put it in another place, the whole is very different. Likewise, different components can create wholes that are (more or less) the same. For instance, think of artifical intelligence. While it is silicon based, it shares many characteristics of the human brain. A Gestaltpsychological example often given is that you can, with different types of units, create the same faces.

On the story:
-how does one, in physical reality, put the pieces together without glue or without adding any other components?
-the problem with the second one is that the parrot is not the same: even if it is a clone, then still he does not have the same memories. furthermore it would take a lot of time to have the same age as the initial parrot.
Similar I believe with the more general 'proof': if you cut something in three, you cannot necessarily reassemble it into the whole it was before, at least with current stance of technology. Say, in your cutting the universe, you cut a human in two. Can you reassemble him? It may depend on the nature of 'cutting', but standardly, cutting involves killing the person – as in making his organs stop to function, blood no longer supplying the organs, and so on. In this case, we cannot 'go back in time' to restore his bodily functions.
Of course, if one would allow going back in time, then there would be no discussion – since the parts can always restore the whole simply by going back in time.

Jeff: David, thank you for the comments!

Regarding your take on the 'whole is more than the sum of its parts', please note that in the blog I tried to make it explicit that holism and emergence are not the same thing. Your comments on the chair, artificial intelligence, etc. are regarding emergence, not holism. This is perhaps another blog itself... ☺

Regarding the physical act of cutting: it is often not necessary to destroy the sytem in order to study its constituents. Think of it more as being able to observe the three pieces of the pie without actually dividing it up. We do this automaticaly, for example we differentiate between the coffee cup and the coffee inside it, without pouring the liquid into another container. I suppose the crux of my argument would be to ask: "at what point in pouring the coffee out of the cup would the coffee+cup system be completely reduced to two separate systems"? This is just another version of one of Zeno's half-of-a-half paradox, but I think it fits.

The joke was meant to be an easy mnemonic device to remember the argument that analysis and holism are fundamentally the same thing. As Dale Carnegie says, "you can't win an argument" so maybe it is not bad to present the blog discussion as a joke, that way both reductionists and holist can at least have a laugh.

In any case, I will work on fleshing it out into a 10+ page paper in the hopes that it makes it into your annual publication.

Thanks for the opportunity to blog at BossensNonFichtion!

Jeff

David: Hey Jeff,
although in private I mentioned the sentence was a bit confusing, I think I did interpret it right, to think that emergence meant the whole is more than the sum of the parts. I was not really providing any counterargument to what you said, just clarifying my position on the emergence issue ☺.

Maybe you can write about the Zeno-paradox in a next blog? Or on the whole and its relation to the sum of parts?
All appreciated highly 😊

David: "The system can be finite or infinite. In our case, we define the system as the universe at large. Now if we were to divide this system into thirds, and then put them back together again, we would find that no matter how hard we tried, we could never distinguish the difference between the sum of the thirds, 0.999 and our original, entire system. "
I think it is unimaginable to divide an infinite thing in three pieces. So I guess finite would be better. Also, 0.999 of something very large will not be the same – but it doesn't matter since we haven't cut any piece of it anyway.
More generally though, what your article attempts to do, if I read it correctly, is to say that the whole=the sum of the parts – which negates emergence, as you have defined it. Personally, I don't think anybody means this – that when we analyze its components we would somehow find something 'more' – when they say the whole is larger than the sum of its constituents.
'The whole is more than the sum of its constituents' usually applies to the fact that different configurations of the same elements will lead to different wholes, with entirely different properties – and same configurations of different elements, will often give more or less analogous wholes. That is, order, or patterns if you will, matter(s). Perhaps develop your intuitions on this further for your article for the Bossens Yearly, and include the Zeno paradox. Be sure to respond to the issues I just mentioned.
Maybe next blog on another topic, or my blog will look somewhat too specialized on this topic, and not be representative of the larger range of topics that our website represents.
Take care

The unbridgeable gap

Ernst Mayr believes there to be an unbridgeable gap between biology and non-biology, believing there are processes that cannot be described in physical-chemical terms.

He also argues that entropy, the increase in disorder, does not apply for biology since it is an *open* system. Libb Thims knows to inform me:

The general issue is that there are two kinds of entropy, one when heat is put into a system, such as to turn an ordered ice cube into disordered water vapor, which seems to be the kind that Helmholtz discussed when in 1882 he described the magnitude of entropy as a measure of disorder.

The second kind of entropy is what is commonly known as "entropy increase", which has to do with the Carnot cycle, namely when a certain amount of heat is added to the system (volume expansion), followed by removal of nearly the same amount of heat (volume contraction), so to bring the system back to its so-called "original state". The first state and the second state will not exactly be the same, the quantifiable difference between the two states is what Clausius calls the equivalence value of uncompensated transformations, symbol N.

This is what is commonly known as the increase in the entropy of the system and is the quantitative measure of change and hence evolution (as Darwin called it) or metamorphosis (as Goethe called it) or as chemical synthesis (as we would call it). The problem is that biologists get the two types of entropy mixed up and fumble up the whole thing into a convoluted verbal mess."

In any way, I don't think there is an unbridgeable gap between the biological and the non-biological, at least not in terms of underlying laws. It seems, by chemical and physical laws, that we humans are governed by fundamental principles that are true for all of the universe. Biology *can* be reduced to chemical and physical explanations.

However, in terms of methods, which is what Mayr refers to, it may be the case that low-level extrapolations upward may not work. Sometimes the behavior of a whole is highly different from what one would expect from adding elements together. Even slight changes in configuration can lead to extremely different results.

The verbal biological method still serves as a useful and efficient method to describe and predict these larger scale systems. Usually this is sufficient, as witnessed by the success of many biological streams of research. Ernst Mayr believes in some kind of schism, with regard to the methods of biology vs those of physics. Complex systems can be best predicted using the historical, verbal method, rather than the mathematical method. A complex system, by definition, is a system composed of interconnected parts that as a whole exhibit one or more properties not obvious from the properties of the individual parts.

Also, evolution as a term cannot be replaced by just synthesis, since synthesis says nothing about the gradual change of genetic material. It may comprise this, but we cannot rely on overarching supercategories all the time – sometimes we need to be very specific.

Jeff:

The unbridgeable gap exists only to those who are not wiling to swim.

Libb:

Yes indeed, our friend Russian physical chemist Georgi Gladyshev (post above) seems to be an example of one who does not wish to swim.

G.Gladyshev:

Libb,
Yes indeed, I am your friend. However, I do not wish to swim (as you would like to swim) in some problems.

Libb: *The term "swim" is a metaphor for those who are afraid to dive into Darwin's "warm pond" model of the origin of life and to see if it actually hold up in modern scientific terms.*

On one hand, you hold fast to the cherished term "life" and on the other hand you hold fast to "thermodynamics". The term life, however, is not a term or concept found in either The Mechanical Theory of Heat (Clausius, 1865) or On the Equilibrium of Heterogeneous Substances (Gibbs, 1876). The reason for this is that heat engines are not alive. Yes, the movement of heat between a hot body and cold body through an intermediate body can cause cyclical movement (you and I writing in this forum being one example), but this movement whether actuated on the piston or pen is not living movement, but simply movement.

What you need to do is to let go of the term "life" or living being as these are not part of modern hard science, but simply residual terms of mythology and religion. It took me many years to grapple with this issue. The emergence view or "mixed evolution" view, as you have called it keeps the mind content only for so long, but eventually admits to the inconsistent view that at one point there was a "sort of alive" molecule that led to the first living molecule (or living being as you call it).

G.Gladyshev:

Libb,
You often talk about modern science and modern notions and terms.
Who said that your notions and terms are modern? The vast majority of physicists, chemists, biologists and other scientists do not agree with many of your concepts and terms.
I am opposed to any eclectic approaches in science!
I always remember the words of Dmitri Mendeleev.
When we investigate the world it is ought to avoid the "three equally destructive extremes:
(1) The utopias of dreams, wishing to comprehend all in one rush of thoughts,
(2) The jealous conservatism, which is connected with the complacency,
(3) The arrogant skepticism which does not wish to stop in any place. "
This is my, maybe, not a good translation.
In Russian:
При познании мира надо бы избегать "трех одинаково губительных крайностей:
(1) утопий мечтательности, желающей постичь все одним порывом мысли,
(2) ревнивой косности, самодовольствующейся обладаемым,
(3) кичливого скептицизма, ни на чем не решающегося остановиться".

Jeff: I think that this whole discussion will soon be over.

Evidence is mounting in the field of synthetic biology (just a fancy term for a specific branch of chemical engineering) that will culminate in the creation of the first synthetic cell, via a series of chemical reactions, performed in a laboratory.

Those at the forefront of this work are probably the research group lead by George Church at Harvard Medical School.

http://news.harvard.edu/gazette/story/2009/03/taking-a-stride-toward-synthetic-life/
http://origins.harvard.edu/ResearchOverview.html

To Prof. Gladyshev: maybe a second edition of your book should be in the works...

To quote the mission statement from their own website:
"Similarly, we do not understand life's origins -- how life emerges from chemistry. We do know that the chemistry of life on Earth is rather restrictive in its molecular permutations."

Libb: That my approach is the "modern approach" is evidenced by the fact that in the history of humanity, the first decade of the 21st century has seen three independent published calculations of the molecular formula of the human (Sterner & Elser, 2002; New Scientist, 2005; Thims, 2007), which "considers whole organisms as if they were single abstract molecules" (Sterner & Elser, 2002), which is a step above the older 20th century view that the first organism was a "polyhierarchic self-organized systems" (Gladyshev, 1978) or a dissipative structure (Prigogine, 1955).

When you actually let this view (human = single abstract molecule) sit in your mind for some time, there are many deductive repercussions that follow, the most revolutionary and difficult of the repercussions being that molecules are not alive. Australian engineer Vangelis Stamatopoulos comments on this (15 Nov 2010):

"A very logical, rational and totally convincing argument indeed. When I first read it, it certainly made me think about my views and understanding of the experience of existence or consciousness (i.e. life?) that I began to doubt my understanding of it."

http://www.atheistnexus.org/profiles/blogs/its-life-jim-but-not-as-we?xg_source=activity

In Stamatopoulos' own words, the atomic or molecular view of humans makes him "doubt" his understanding of the concept of life.

Your continued name calling at me (dreamer, jealous, arrogant, etc.) only works to shed light on the insecurities in your own theory.

David :

Hi Jeff, has been a while.
"Evidence is mounting in the field of synthetic biology (just a fancy term for a specific branch of chemical engineering) that will culminate in the creation of the first synthetic cell, via a series of chemical reactions, performed in a laboratory."
The above link was an example how ribosomes can make proteins. Yes artificial cells have been made, it wasn't this one however. But anyway: what do you think this means in terms of the life debate? Not much, since we already know that we are composed of chemistry. 'life' is just the denominator for things that have DNA to code for proteins. In terms of the unbridgeable gap, I do not see what this proves? I don't think anybody believes that 'life' is not some chemical reaction? the unbridgeable gap, correct me if I'm wrong, refers to that the methodology for studying the behavior of biological systems is often of a historical nature, whereas other things are for example studied by mathematics or chemistry.

Jeff: Hi D. Boss, my answer has also been a while in the making!

For me the fundamental question regarding biology and its interaction with chemistry and physics is that the physics of microscopic systems is described and defined by interactions whose rules are difficult to decipher for macroscopic systems consisting of large number counts of microscopic entities. If you are interested in these things, please read Howard Pattee and Ernst Mayr for very different but equally relevant takes on the fundamental issues behind biology and physics being considered as separate fields of scientific inquiry.

There is little doubt in my mind that ecosystems follow the same laws of physics as helium atoms, however we are still lacking a well-tested theoretical framework which is united both in phenomenology and theory for biological/ecological systems. Also, it is not a bad idea to look at "old school" ideas such as the De rerum natura for inspiration. If you are interested in Goethean science, I would recommend looking up "zarte Empirie" in Google for many good sources of his holistic perspective.

So far, I would put my money on a mixture of constructal theory, hierarchical thermodynamics and some of Libb's human molecule concepts to describe macroscale configuration, although I am still not sure if it is not fundamentally the same as looking at the path-dependency of free energy functionals in a well-defined phase space. Gradients drive all macroscopic physical processes, it is really a question of the measurement and correlation of these driving forces behind macroscopic systems which makes them difficult to comprehend.

These are deep scientific questions which still require inquiry. Good luck!

Regarding your assertion that DNA is the least common denominator in defining chemical "life", I would disagree. DNA is a microscopic configuration of matter which serves concomitantly as code and program. It requires feedback and interaction across both micro and macroscales. The crossover between micro and macro even after the advent of statistical mechanics is still an unresolved (but largely ignored) area in modern physics. Fundamentally, there appears to be some difference when we study individual events vs. average values. Some of the questions to be answered when investigating this crossover are:

Are macroscopic systems truly stochastic? What is the fundamental cause of the "elementare Unordnung" in microscopic systems? Is the second law truly an expression of probability?

Since DNA acts across these scales and since we do not have a physics principle which adequately describes this transition, I would be very careful in stating that DNA is the least common denominator for the commonplace notion of life...

Once again, I would refer you to reading Howard Pattee for more consideration of why we should not stop at DNA when asking "what is life?"

The life-debate revisited

With regard to our life-debate, it is interesting to note that 'life' on other planets as we would call it, would not have DNA - most likely. So perhaps, to make clear distinctions between 'life-forms' with DNA vs those with other types of genes, we perhaps can propose DNA-centric molecules vs whatever-centric molecules for the other types. My long argument still holds, but when opposed to alien 'life-forms' which do not have DNA, we might conclude that it is more practical to define it as DNA (or RNA) -centric molecules, rather than 'bio'. Something Libb had proposed in response to my DNA-propoal.

Some time, we will see other genetic material than DNA/RNA, and other types of cells, or none at all in what most would consider 'life forms'– most likely on other planets. Although we in that case can still keep the term bio, this term stems from a root that has nothing to do with 'DNA-molecule' and it will be somewhat asymmetric to call the one bio and the other some name related to its gene-molecule. I'm still not sure what the word cell would mean in other 'life forms'. One could say that 'genes' are universal for biology – even in alien civilizations.

Given that we will (perhaps?) ultimately view intelligent robots as 'life forms' as well, cells also seem not to matter - although most would then say the 'non-biological organism' , 'cyborg', or 'silicon based life-form'. It seems a criterion as intelligence seems to matter.

It would be indeed difficult to recognize 'life', at least on other planets. The only thing we might consider to be alive (if my assumption of is that which moves as if it has certain expectations. What seems to matter is intelligence and movement. For instance, if something moves away whenever we take a step closer, we might consider it intelligent and give it the name 'living'.

There are many options of course. We might recognize other patterns in aliens which we also might term cells, and other molecules that seem to be necessary for creating those cells. We cannot know, but for now, we are still comfortable with the following propositions in the life debate:

1. If equating life with 'having a soul', then life is defunct.

2. 'biology' can be used to indicate systems that rely on DNA / RNA for protein synthesis and cell division.

3. If, hypothetically, we would find alien entities which do not have DNA, but do have some sort of 'regeneration molecule' and/or 'protein synthesis molecule', then we consider the term 'biology' or 'genetically-based molecules' as the superset, and as a subset for instance, we can, in the case of DNA as regeneration and protein synthesis molecule, take 'DNA-centric molecules'.

4. If we consider other 'life-forms' without such regeneration or protein synthesis molecules, we might question whether it is intelligence that matters in the life debate. Indeed, it seems intelligence, in any case, is at least a peripheral characteristic of life. Given that, quote Michio Kaku "the smartest robot is as smart as a lobotomized retarded cockroach", we might consider intelligence as a condition to *perceive* something to be alive. Perhaps if robots would be highly intelligent, we would call them alive. Depending on which terminology we use by then, in most definitions this would be wrong, since it has no soul, no genes, no cells. It does have intelligence however. One may try to define intelligence via physical principles, to make a distinguishment between intelligent and not-intelligent. However, since intelligence, according to the digital philosophy, is but a relative contrast, intelligence, is a quality that comes in different quantities. In other words, the question is not whether something is intelligent, but rather to what extent it is intelligent.

5. If we consider the debate of aging, it becomes clear that there are some thermodynamic methods to establish whether something is of anti-aging value, it seems that aging can be anchored in a thermodynamic definition. Since aging can be defined as 'coming closer to death' or 'becoming less *vital*', perhaps this debate is very relevant to the life debate.

Aging and Life

Some questions remain unanswered, such as aging and its apparent relation to thermodynamics – see Gladyshev's anti-aging formula.

David: *If aging is able to be anchored in thermodynamics, doesn't that mean that 'life' vs 'death' can be anchored in thermodynamics?*

Libb: *You're trying to anthropomorphize everything. This is what's called the top down approach. A biased way of doing things. The correct unbiased method is called the extrapolate up approach.*

To exemplify, we know that the "half-life", a term coined by English physicist Ernest Rutherford in 1890 to describe the radioactive decay of radium (1620 years), of carbon-14 is 5730±40 years. Carbon-14 decays into nitrogen-14 through beta decay, releasing an electron and an electron antineutrino in the process:

$C14 \rightarrow N14 + e + v$

in which a "stable" (non-radioactive) isotope (N14) results. This is no different, scaling issues aside, then when un unstable marriage MF decays or has a "break up" (divorce) into single "stable" products:

$MF \rightarrow M + F$

The "half-life" for marriages being about 16 years.

Rutherford, here, did not think of this 5,000-year C14 decay reaction process in terms of life, death, and aging, as you seem to be doing. Yes the carbon-14 decay process is governed by the first and second law of thermodynamics, i.e. energy conservation and entropy increase; but from this fact we do not conclude that the "aging", "life", and "death" of the carbon atom can be "anchored in thermodynamics". Again, what you are attempting to do is force mythology into chemistry and physics, which only leads to absurdities, e.g. Christian de Quincey, speaking about "miracles", such as on the extrapolate down page.

David: nevertheless, 'half-life' is only one word away from 'life', so at least Rutherford was slightly thinking about the word 'life'.

in the 'life' vs 'death' debate - or how would you call both respective terms - personally I think in terms of biological degradation caused by the end of cell division because telomeres stop functioning, or of course insufficient blood flow, or cancer, which causes some vital functions to stop - a process of starting to decohere and consequently biological degradation (getting eaten by worms and such).

'Aging' is an anthropocentric term of course, but maybe, and this was my question to you, there is some thermodynamic process behind it (yes, Gladyshev and you should know about this, since you have found anti-aging benefits by means of a thermodynamical equation), which would entail being able to give a scientific definition of 'aging', which would consequently allow a less anthropocentric term for 'aging' - the transition from 'life' to 'death' : yes those concepts 'life' and 'death' may then also be redefined in a scientific definition. As said, telomeres are a good marker for human aging. Yes, even anthropocentric concepts can be studied - if not, why do you and Gladyshev find methods to counter aging? It is because these anthropocentric terms should be more scientifically defined that I ask this question. we may replace or equate anthropocentric terms then with more correct terms, to gain better understanding what it is that we want to avoid. [aging and death]
Or was it a suggestion that the human molecule, similar to C-14, is subject to radio-active decay, which would equate with 'aging'?

one such possibility of "life anchored in thermodynamics" is the reversal of the second law
"Reversals of the second law are a regular phenomenon, and [they are identified] with what is generally known as life."

The latter is a quote from Sidis, who conjectured that life is a reversal of the second law of thermodynamics. Interesting that Libb has largely ignored this in his 'life = defunct' debate. Also interesting is that Libb seems to advocate a bottom-up approach, while it should be clear that predictions at many levels can be made. *Should* Libb do chemistry by means of string theory? Sometimes maybe, but if one tries this for every prediction, your science will be severely retarded. *Does* Libb do chemistry by means of string theory? No. It is clear that different methods will be fit for different levels of prediction. Even if thermodynamics is somehow a *primary principle*, which is suggested by many to be so, then still, it seems obvious that many other branches of science have been successful without directly using it- different scales or study objects require different methods, although it is often wise to connect the methods with each other in a large overlapping framework.

We can be sure that we have a heart, muscles, veins and telomeres. Should we then ignore these, simply to do everything via chemistry? Good luck in describing what's wrong with a patient who has a heart attack, and predicting what will happen. Mmh, the human molecule should remain the same, since we don't add any chemicals, and we don't throw away any chemicals. How strikingly important then, for medical science for instance, is it to have macro-descriptions and hence macro-predictions. Or even the following example: shoot somebody in the head; notice that he dies. Why so? His brain stopped functioning.

Also the famous Tesla quote implies a certain definition of 'life'. To debunk a concept, is to have a certain definition of it. This definition is a mythological one- in terms of true free will, a soul. But the definition does not need to be a mythological one. Many biologists have a reasonable definition for it, nowhere mentioning free will or soul. The bottom-up approach can only yield predictions on some scales, or some domains.

Cognitive factors in 'love' for instance may be considered. It may be that they are secondary – that some chemistry is needed first ... However, consider arranged marriages. No chemistry need be involved with mating. Try to describe arranged marriage in terms of chemistry and know that you are doing useless work .

One can disagree with the statement 'to debunk a concept is to have a certain definition of it', by saying 'there is no definition for it, that's the problem'. The first problem is that both Thims and Tesla see life as equivalent with 'having a soul' or some special physical life force. In this case, yes life is defunct – but, clearly, there is a certain definition of the term. The second problem, with the fact that there would be no definition, is the fact that life can be equated with 'RNA-centric molecule' or 'systems that use DNA/RNA to synthesize proteins and divide its cells'. Other peripheral characteristics of life exist, but these are the main characteristics.

Libb: *"why do you and Gladyshev find methods to counter aging", "we" didn't find anything. Gladyshev has an anti-aging formula, not me. The aging page is here:*

http://www.eoht.info/page/aging

That is what I do here at Hmolpedia, find thermodynamics-related human theories collect that theories and points of view in form of some 2,600+ articles. The entropy "reversal/reduction" articles are here:

http://www.eoht.info/page/Entropy+reversal
http://www.eoht.info/page/Entropy+reduction
http://www.eoht.info/page/Local+entropy+decrease

You're barking up the wrong tree, if you are attempting to preach a "life theory" to me. The three things JHT will no longer publish are (a) god theories, (b) information theory, and (c) life theories. This is similar to patent offices who since circa 1900 no longer accept perpetual motion theories, of which (c) is an example.

Re: "Or was it a suggestion that the human molecule, similar to C-14, is subject to radio-active decay, which would equate with 'aging'?", you need to stick with neutral terminology, e.g. synthesis/analysis:

http://www.eoht.info/page/synthesis

What were aiming to do here, as stated on the main page, is to explain human movement quantified by: heat, work, energy, entropy, Gibbs free energy, activation energy, coupling, irreversibility, extent of reaction, bond energy, spin, and other factors, tending to be quantified as conjugate variable pairs.

When you or a rock move through a distance by the action of a force, we say that work is done; we do not, however, say that either is alive.

The same is the case with reaction rate and time of existence. We can speed up or slow down a reaction by adding heat, adding catalyst, or via modifying the substrate, i.e. add time (age) to the molecular synthesis (putting together) or analysis (taking apart).

Doing science that is in our advantage – why anthropocentrism and top-down approaches are necessary

David : *clearly it makes sense to research things from a top-down approach since we are all interested in living long and healthy - as is also the case with you, cooperating in marketing the anti-aging thing (and I'm glad you did or are, or whatever). it is interesting to start from a antropocentric goal - for instance creating a device for seeing (light bulb), or transporting (a car), or against 'heart attacks' (pace maker). I hope you agree that all science come from anthropocentric goals - preservation of human race, 'fun', 'anti-aging'. Is 'aging' another term that does not exist? sure, we can find neutral terms for it, if we can reveal all the factors in 'aging'- a goal of research done by many, for which I am very glad. in that case, aging would have a definition related to all those factors. enumerating all these factors then is simply too convolusive and unnecessary since we know what we are talking about. without top-down approach, we wouldn't be studying how to live long. your objection to 'aging' is that it stems from a top-down method. however, studying from a bottom-up approach will never lead to a cure for aging - if one can't acknowledge that the term aging exists and hence one may not try to define it, one cannot research it.*
things top-down approaches can predict with absolute certainty for instance: remove the heart and blood flow will be disturbed, resulting in not supplying oxygen and such to the cells. not recognizing things like 'heart', 'blood' and 'cells', because they are not defined in terms of thermodynamics is retarding any medical conversation.
One can define them in chemical terms of course, and in some cases this is useful or more predictive.

I understand that you as well take a top-down approach sometimes - how else could one achieve their goals- by trying to 'give a definition of love', or by attempting 'a scientific description of death'.

Yes, science starting from top-down terms has achieved much. 'What is mental illness?' 'What is love?' 'What is aging? How can we avoid it?' 'What is the function of the prefrontal cortex?' Etc…

There are many things we know with absolute certainty on the macro-level, so the argument that "it's wrong because it's not defined on the bottom scale", is a false argument. Sure, in some instances, micro-descriptions lead to better predictions, but often these micro-descriptions are unnecessary for giving predictions at the macro-level. Should we use string theory to determine the location of my penis when a naked woman stands in front of me? Most likely, neuroscience and biology will give accurate predictions. Cut a certain specialized area (the hypothalamus and amygdala for instance) and the in the brain, or some artery, and the erection will not occur. Inject testosterone or Viagra and know that erection will occur. Indeed many of these macro-level predictions are highly useful for anthropocentric goals. Indeed, different methods are useful at different levels, and serve different goals.

In this light, 'life', 'love', 'aging' and others can all be defined scientifically – although this would necessarily entail cutting some connotations of the term, at least in scientific conversations. For instance, with 'life', no physical life force or vis vitalis can be kept in our definition of 'life'.

Truth of the matter is that Libb himself uses anthropocentric concepts. He even made a paper termed 'Thermodynamical proof that good triumphs over evil'. In that paper he substituted 'good' with 'natural' and then 'natural' with 'dG<0', and the opposite for 'bad'. I personally have no problems with this. Definitions are specified, so we can talk. But Libb apparently either has made a shift in philosophy since 2011 (now a year later that I'm writing this) , or he is inconsistent. Indeed, words can be equated with things that it didn't equate with before. This in fact is what much of the 'life debate' is about: Libb equated with life something else than most biologists – some physical law present in the atoms of living things, a soul, or a life force. I consequently adopted this definition of life, but then made another definition for biological organisms – a definition more relevant to what the discipline biology actually studies – even if it were only to make clear what we were talking about.

The bottom-up approach is not necessarily better: if one small mistake exists in our understanding of the 'bottom', then any inference to higher levels is wrong. Most bottom-up approaches only go one level up, hence the saying 'biology is applied chemistry' for example. Furthermore, if we look at quantum-mechanics, we cannot use the bottom-up approach and say 'we know this for sure, so it must apply to larger scale systems as well': mechanics of large systems are very different! Conversely, if we look at macrostructures directly, we don't need perfect knowledge of the bottom, and we don't need the viability of extrapolation. For instance, when the heart is to be analyzed in terms of chemical components, if we try to derive some prediction of the larger whole (the heart), then this analysis should be perfect, and every micro-conclusion must be translated correctly into macro-conclusions, assuming that the micro-conclusion was correct. However, when we say that 'without the heart, blood stops flowing to the cells', we have a certainty that exists at the macroscopic level. In this case, we have not used a microscopic analysis to extrapolate predictions at the macro-level, yet we have a 100% certainty. In sum, the bottom-up approach is only useful and efficient in some cases; the top-without bottom approach and the top-bottom approach are useful in other cases.

On Schrödinger and Sidis

David: *Is this proposition of Schrodinger "What an organism feeds upon is negative entropy ", roughly the same as the proposition of Sidis that "Reversals of the second law are a regular phenomenon, and [they are identified] with what is generally known as life." ?*
can you give reasons for this being false? also, I'm not sure where to ask, but I was wondering on some other of your philosophical stances, besides 'life' and atheism- am writing about it.

Apparently, this definition got crushed by others in the coming decades.

Libb: *Re: "Is this proposition of Schrodinger "What an organism feeds upon is negative entropy ", roughly the same as the proposition of Sidis that "Reversals of the second law are a regular* phenomenon, and [identified] with what is generally known as life."?, not exactly. Schrodinger started from Ludwig Boltzmann's gas theory view, Sidis from William Thomson dissipation of energy views. Beyond this, Sidis' argument is more complex, details aside.

The gist of both of their attempts at solution, however, is captured well in Schrodinger's appended "Note to Chapter 6":

http://www.eoht.info/page/Note+to+Chapter+6

wherein you see how in the years to follow he gets ripped apart by hardened thinkers such as English physical chemist and chemical thermodynamicist John Butler (1946) and American chemical engineer Linus Pauling (1989). In the end, Schrodinger recants that future investigators should "let the discussion turn on free energy". Let this advice soak into your head.

Instead of thinking about entropy, second law, life, etc., think about your existence, as a moving molecule, bound in various states of existence, some more or less attached (in complex interpersonal relationships), that change as time progresses.

The thermodynamic potential that quantifies each of these states, energetically, is "free energy", not entropy:

http://www.eoht.info/page/Thermodynamic+potential

In this framework, you can begin to think of your various "states" of existence in terms of "integrals of forces with respect to distances" (Lagrange):

http://www.eoht.info/page/Potential

All of this is summarized, in modern terms, on thermodynamic data tables, wherein, as shown in the following link:

http://www.eoht.info/page/Affinity+table

each "molecule" (human or otherwise), in a given state of existence, is quantified by a free energy value.

I now show the reader Pauling's criticism of the 'life feeds on negative entropy' – to be found on the above link on 'chapter 6'.

In his 1989 memorial chapter "Schrodinger's Contribution to Chemistry and Biology", American chemical engineer Linus Pauling (1901-1994) probably gives one of the harshest critiques of Schrodinger's overall ideas on the thermodynamics of life. Pauling, after telling how much respect he had for Schrodinger for his work in quantum mechanics (particularly for his Schrodinger equation), begins his rip into Schrodinger's What is Life? lecture with the following mock:

"In his discussion of 'negative entropy' in relation to life, he made a negative contribution."

Pauling, then, after quoting Schrodinger's two main excerpts about "keeping aloof" and "feeding on negative entropy", continues:

"Schrodinger's discussion of thermodynamics is vague and superficial to an extent that should not be tolerated even in a popular lecture. In the discussion of thermodynamic quantities it is important to define the system. When he is writing about a change in entropy of the system, Schrodinger never even defines the system. Sometimes he seems to consider that the system is a living organism with no interaction whatever with the environment; and sometimes it is a living organism in thermal equilibrium with the environment; and sometimes it is the living organism plus the environment, that is the universe as a whole."

So, shifting from one system to the other, Schrödinger made his argument – albeit temporarily – seem like correct.

I also show the reader what types of thermodynamic potentials exist. (again to be found on Hmolpedia under 'thermodynamic potential'.

In thermodynamics, **thermodynamic potential** is
the name given to a function whose minimum
gives the equilibrium state of a system subject to specific constraints. [1] Among the most often encountered thermodynamic potentials cited include:

- Negentropy (isolated system)
- Internal energy (quantities of extensity constant)
- Helmholtz free energy (temperature, volume, and amount of substance constant)
- Gibbs free energy (temperature, pressure, and amount of substance constant)
- Enthalpy (entropy, pressure, and amount of substance constant)

Stated verbally, the conception of thermodynamic potential provides for a description of the direction of evolution of physical systems. Through the second law, the science of thermodynamics states that a system evolves in the direction that minimizes an appropriate thermodynamic potential, for example the "negative of entropy" (neg-entropy) for isolated systems, or the Gibbs free energy at constant pressure and temperature. [2]

Chemical reactions
The application of the thermodynamic concept of potential to the kinetics of chemical reactions (transition state theory) provides a criterion for selecting the optimal pathway for a transition, usually the pathway with the transition state of the lowest free energy. **In this perspective, by providing a direction for systems to evolve and an optimal pathway, thermodynamics offers a way for answering *why* things happen the way they do.** [2]

The latter (my bold) seems an interesting fact to reflect on.

To go back to the conversation of life, it seems negative entropy is possible but only in isolated systems.

David: *the main criticism of pauling lies in that shrödinger haphazardly shifts definition of the system. I have read that neg-entropy is possible, but only in an isolated system. Do we know of any truly isolated system or are they just theoretical constructs?*

Unsure whether life really 'feeds on negative entropy', it is still the case of course that there are clearly distinguishable characteristics in life – DNA , RNA, cells and such. If indeed it shows that there is no physical principle that would be relatively special for life, I would not be surprised.

I will repeat what I said in the reconciliation chapter: In any event, we need to acknowledge that Libb's view holds true in many ways, but we need other ways of looking at the world as well, if we want to create a better future for mankind – see the discussion on anthropocentric goals.

The conversation continues:

Libb: Re: *"the main criticism of pauling lies in that shrödinger haphazardly shifts definition of the system"*, to correct you, the main criticism of Pauling is that Schrodinger is using the work of Boltzmann (gas theory) to explain animate existence instead of. in his own words, the "great work by J. Willard Gibbs on chemical thermodynamics" (chemical theory). Human movement and ordered existence is explained by the 700-equations of the latter, not the one approximate equation of the former.

Re: *"I have read that neg-entropy is possible"*, this statement doesn't make any sense, please provide reference.

In any event, to put you on the right track, entropies for all ordered structures (chemicals, molecules, humans) are, by definition, either "positive" (no pun intended) or zero:

http://www.eoht.info/page/Free+energy+table

You, in a given state, will have a measurable "positive" value of entropy, e.g. 250 kilojoules per kelvin per hmol, say in year 21 of your reactive existence.

David:" " *I have read that neg-entropy is possible"*, this statement doesn't make any sense, please provide reference."
it was on your 'thermodynamic potentials' page. Or what does one mean with Negentropy (isolated system) ?

so the one is that he should use Gibbs free energy instead of entropy, and the second is of the type: even if we use neg-entropy as a concept, then shrodinger has used his systems inconsistently ? thanks

Libb: The gist of the Schrodinger "What is Life?" lecture, is that he was giving a lecture to a lay audience; resultantly he ended up dumbing it down to the point of becoming a sloppy derivation (hence the attack by Pauling, and others), at least for the thermodynamics part.

The second law (entropy tends to a maximum) can be stated negatively. So, stated negatively, in terms of thermodynamic potentials: neg-entropy tends to a minimum – the amount of thermal energy left for useful work is minimal. My statement 'neg-entropy' is possible is a bit confusing – let's just say that if I implied that reversals of the second law were possible, that I would most likely be wrong- unless there was some external work done to achieve this in some isolated system. As is the case with a refrigerator, where heat flows from cold to hot – and, as may be the case with 'life'?

Schrödinger in his own words referred to that he should have conversed about free energy, but found that it probably was too difficult for the lay audience.

A Hmolpedian debate outside of the Hmolpedia

An interesting debate to illustrate some of my disagreement with Libb is the following conversation on http://www.thescienceforum.com/biology/26072-why-cant-science-make-life.html.

4n4nd: *Science has advanced this much. Still we are not able to create a life in the lab. Of course we did cloning. But why cant we restore the life of a dead man?I want to know what's the factor that hinders us?*

First of all, we *are* able to create life in the lab, it has been demonstrated by using synthetic cells. Second, theoretically speaking there are two different interpretations to the question 'why can't we restore the life of a dead man?!'. The first interpretation is more loose, meaning that we want to have the same consciousness regardless of composition. This, in the future will be possible with mind-uploading, where for instance slices of the cerebral cortex are decoded into computer circuitry, and hence we get artificial brains with the possibility of backing up our 'mind-file'. The second interpretation is that we, either by replacing some organs, or by reviving our heart or something, revive a living person in more or less the same state. In one way this is already possible with defibrillation, so to get the heart pumping again. But when the heart definitely stops pumping it is more difficult, though not impossible – at least in theory. The factor that hinders us is that nanotechnology is not yet sufficiently advanced to prevent problems in the body. Nanotechnology will give immense benefits in the realm of prevention of death.

However, suppose that we are dead and that nanotech did not prevent it from happening. Could we revive a person from death? The biggest problem is that blood stopped circulating around the cells to provide them with nutrients, so most of the cells have died off. If we however use cryopreservation, a technology yet to be perfected, we can forestall the degradation process, to gain time – until some discovery has been made to restore the functions of the body of some dead person.

Libb Thims : *The reason why "science can't create life" is because "life" is something that does not exist: it is a mythological construct handed down to us from Egyptian mythology, through our childhood and cultural religious-based teachings, but is not in fact something that is found existent in the hard sciences: chemistry, physics, and thermodynamics. This line of reasoning (quite refreshing to say the least) is what is called the "defunct theory of life" (Google), first positioned by Tesla, but only recently come into light with rise of the thermodynamic analysis of life, aka Schrodinger and the feeding on negative entropy description.*

The debate is long and winded, but the nuts and bolts are the fact that a human is an animate 26-element molecule (Ecological Stoichiometry, 2002; Human Chemistry, 2007) and molecules are not alive, but only can be viewed as having increased levels of animation and reactivity. In short, the human evolved over time from hydrogen atom precursors. The hydrogen atom is not alive and neither is a human, which is the reason why, since the time of Faust and his homunculus (laboratory-created life thing) to the present, science cannot produce it: because it is a "forced idea" (mythology forced into chemistry and physics).

Barbi: *Well then science should have no problem making a animate 26-element molecule*

Indeed, this reminds us of the Gestalt-conversation. Can we synthesize the 26-element human molecule, then apply solar force on it to animate it and consequently say 'this is a human' ?

Wrong. Most likely, synthesizing the human molecule will yield a clump of matter which is not in any way related to a human. It is mostly the DNA/RNA that matters to give a human. And yes, indeed, we can make life – not only by sexual reproduction, but also by test tube babies, clones, and of course synthetic cells.

The first question first: if 'life does not exist', does that mean we cannot make it? Well, it is clear that life, according to the OP, *does* exist, so, if one has a slight interpersonal connection, one would acknowledge that the OP most likely has things in mind that are unanimously called 'biological', such as bacteria, cells. Consequently, the question is not difficult, and we *can* 'create life' – this has been demonstrated.

Furthermore, what proposal is made to revive a person from death, via this proposal of Libb? Sure, Libb may say there is no proposal needed, because the concept does not exist. However, when he looks deep into his heart (figuratively of course), he will acknowledge that he doesn't want to die or 'degrade'. As said, anthropocentric goals drive good science. If not, we would all be performing research that has no relevance to our goals. Yes, science should combat 'aging', should try to avoid 'overexploitation' of the earth, should be 'morally' employed, should be made for 'useful applications' – all terms which are not defined from a bottom-up perspective, but rather, which are defined from our perceptions. However, according to the anti-utilitarian vision of Libb, we should only research 'truth'. Well, certainly truth is important, but how about the truth on how to stay alive? How about the truth that if you're not alive, thinking, breathing – however you want to label it – you cannot do *any* research? How about the truth that if we stopped doing useful things, we would end up not finding truth, because of the demise of society?

I would find it problematic if people adhered to Libb's vision and consequently said 'life does not exist, so you cannot research it'. This way, every debate with a concept that does not start with the bottom, is inexistent. Or at a minimum, every debate about ending aging, saving people's lives, is bound to be stopped by some Thimsian follower – should those exist. Consequently all attempts at investigation of what clearly makes us happy – not degrading, dying – will be stopped, since the concept of dying would be inexistent. Oh yes, even 'happy' does not exist.

Even if I would agree with Libb that 'life does not exist', then still: surely, we can agree that when we are having a conversation to revive someone to 'life'=the state that he was in before, one cannot say that 'we cannot revive someone to that state, because that state does not exist'. So, in any case, the statement that life does not exist is irrelevant, since we all know what the person means when he is referring to with 'reviving an old man to life'.

This is not to say that life is something particularly special in chemistry – it is merely some structure with DNA/RNA coding for its proteins, which has cells. But it is special to us, and so we should find ways to stabilize this structure, and stop it from degrading. Unless of course one prefers to be unstable and die.

4. Intelligence, consciousness and free will

Retinal model of free will and the brain as a hydrocarbon molecule

David: *if one was to look at alien civilizations, one would probably conclude that they have no DNA - am not sure of course. In that light, 'DNA-centric molecule' becomes a more useful categorization than 'biological'. Nevertheless, life apparently indicates 'intelligence', since when we find something intelligent, we will probably call it 'life-form', regardless of having cells or DNA - is my guess of course, I don't know for sure that aliens do not have DNA or cells. So my main question, how would you define intelligence? do you think intelligence is in atoms? or does intelligence relate to the possibility of sensoric input? can this intelligence perhaps be a marker for 'life', why not? also, how would you view anti-aging models such as the one of Gladyshev in this context?*

Libb: Here's a picture of a simple example of an intelligent hydrocarbon "brain" in an animate molecule (ABC model section). [links to the article following]

The human brain is similar to this, comprised mostly of hydrocarbon atoms (EPA and DHA); certainly more complex, but nevertheless a hydrocarbon brain.

ABC model of will | Retinal model of molecular choice

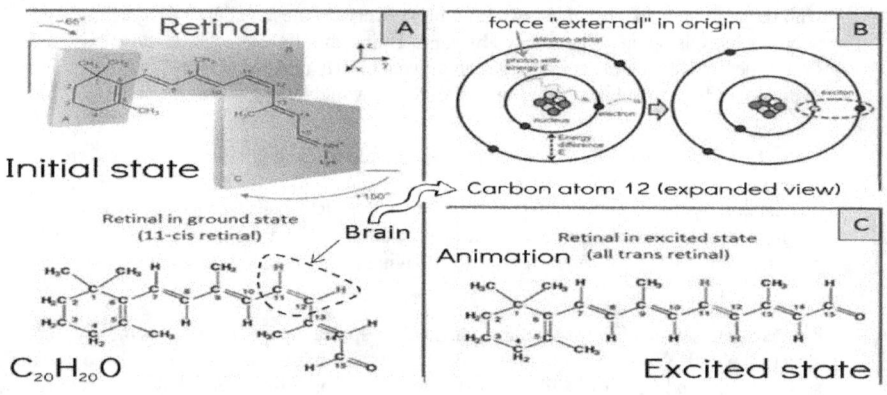

The "ABC model" of **free will**: (A) retinal molecule in ground state (normal state); (B) light (or one or more photons) with a frequency of 400 to 700 nm absorbs into the the carbon atom (note: atom shown is actually beryllium) at the 11 position, thus causing (exchange force) an electron to jump up in orbital position (excited state); (C) the retinal molecule reacts to this by "moving" to the straightened position, a short-lived heightened energy configuration. [14]

The basic model for the description of "free will" or induced movement in molecular life (animate activity), is the movement dynamics involved in the life of the 3-element retinal molecule $C_{20}H_{28}O$ (pictured above), a light sensitive molecule found in the retina of the eye. If the energy contained in a single photon is of the correct wavelength, between 400 and 700 nm, it will function to break what are called pi-bonds found between the eleventh and twelfth carbon atoms near the kink in the structure of the retinal molecule. When these pi-bonds break, this 'forces' the retinal molecule to rearrange into a straightened configuration.

This basic model, in which a molecule is forced to react, i.e. moves dynamically, to a photon or field particle stimulus, is the basic model (poster child) for human molecular life, i.e. for all human behavior. The human molecule, a 26-element molecule, is no different than the retinal molecule, a 3-element molecule. Human chemical reactions will always be exact and repetitive, similar to the bending and straightening actions of the simple retinal molecule. More to the point, the central nervous system of the retinal molecule is no different, complexity aside, than the central nervous system of the human molecule: each CNS is comprised, at its core, of valence shell electron-photon interactions.

This photon inducing, exchange force, retinal-bending mechanism, to note, was is an expansion of the 1913 Bohr model of the atom applied to the phenomenon of molecular movement and mechanism with light interaction as discovered in 1958 by the American biochemist George Wald and his co-workers; work for which Wald won a share of the 1967 Nobel Prize in Physiology or Medicine with Haldan Keffer Hartline and Ragnar Granit. [2]

It seems 'intelligence' relates to having sensors that are hydrocarbon. Unfortunately, hydrocarbon is not a very distinctive property. Neither do all intelligences consist of EPA/DHA – see again artificial intelligence. Libb has not delved deeply in the question of intelligence: although he has given some substances that are related to intelligence, which I'm of course very grateful for, he somewhat avoids the question of computation.

According to my theory, a sensor is merely a passing of a chemical signal, from one molecule to another. There is computation necessary for the processing of this chemical signal, but in the digital philosophy, everything computes or better, everything holds information, which can be processed in different ways by different molecules. This in an attempt to bridge 'information' and 'physics': a molecule possesses information, related to its physical properties. These physical properties can be encoded differently by different coding mechanisms. Coding mechanisms, again, are just molecules which react in some way or another with that first molecule. Quantum computers share similarities with non-deterministic and probabilistic computers, like the ability to be in more than one state simultaneously. This allows for a parallel processing as is the case with the human brain. Consequently, quantum mechanics is crucial for understanding the process of computing, and, in particular, parallel computing. A bridge between the physical and the computational, is crucial to have an understanding of 'thoughts'. In any event, 'thoughts' is not exclusive to biology – depending on how thoughts are defined: if thoughts are defined as verbal or visual representations, then it is quite exclusive to the smartest of biological organisms, and some AI's. If thoughts are defined as holding information, with the possibility of being decoded, then thoughts are omnipresent. The latter term information merely indicates 'stuff happening'. Stuff happening refers to molecular processes such as shown in the retinal model of 'free will'.

Consciousness and free will

Another anthropocentric concept attacked by Libb is 'consciousness'. Interestingly, we note that Libb equates (in most, but not all cases) 'consciousness' = 'free will' = 'life' = 'soul'. At the Hmolpedia on 'animate matter', we find:

For instance, Ubbelohde, however, footnoted his definition of life as animate matter by subdividing animate matter into two classes: (a) non-rational animate matter, in which selection processes are unconscious, and (b) rational animate matter, in which selection processes are conscious. This latter addendum, however, is still soaked with anthropomorphism, with the implicit assumption that certain types of moving matter have consciousness and that consciousness is something that exists in the definitions of the modern physical sciences. This dichotomy can easily be shown to be fallacious by asking whether or not the animate light-induced 'straightening' 3-element molecule retinal is conscious or rather if it 'selects to choose to straighten consciously or unconsciously', which invariably leads to the conclusion that the premise of consciousness is an outdated model that is defunct in a modern physical science or hmol science perspective.

See how consciousness here is equated with free will, which is equated with life and soul in other places where Libb mentions these – albeit most often implicitly. Consciousness, however, need not be defined this way. Consciousness, in many cases, refers to the integration of external or internal stimuli, by means of sensory input. Consciousness then is not grounded on some separate physical principle as both Ubbelohde and his opposition imply, but is merely some macroscopic property that stems from the possession of sensors– no free will involved. We all know what sensors are: without it, we would not experience light, sound, pain, etc.

This again illustrates how I both agree and disagree with Libb on such debates. On the one hand, there is no separate physical law – on the other hand, one can look at the macro-world and see that there clearly are differences because there are different structures, both with different configurations and with different elements, resulting in different properties on the macro-levels. Some configurations, albeit that they have the same function (sensors), can be very different. This does not deny that there is a macroscopic property present that is not present in other properties. If consciousness is merely a epiphenomenon of sensory integration in neural pathways, and sensory input is merely an epiphenomenon of chemical processes, then we can still say this macroscopic property exists.

In this definition, consciousness results in thoughts, again something that is difficult to define from a microscopic perspective, but which clearly exists. If not, I wouldn't be writing this. One itching problem in the approach that 'there has no physical principle been shown, so it doesn't exist', is that we haven't found every physical principle yet. Another such problem is, as hinted above, that there is a long chain of events from the microscopic to the macroscopic, so that extrapolation is difficult – and hence the task seems hopeless, so some might conclude 'thoughts do not exist' , 'love does not exist', 'life does not exist', 'consciousness does not exist', etc. It is not because something is difficult to define that it does not exist – especially given that we are not omniscient. In most cases, as with thoughts and consciousness and love, the search for the definition equals the exploration of its factors. Even if consciousness merely results from chemical interactions, then we can define consciousness as a result from those specific interactions, and of course also how it is apparent at the macroscopic level. Again, for anthropocentric reasons, it is necessary to research consciousness, for instance to allow people to regain it when in coma.

By the way, what we know of the microscopic level, is also merely appearance. We make extrapolations based on superficial observations – this is what science does. In the case with free will, appearance is only appearance – of course, it depends on the definition. In the case of height, smell, etc., there are physical referents. However, because the study object is rather restricted, we have a higher accuracy at the small level than at the larger level. In this way, the bottom-up approach is largely superior – one might argue. This is not entirely true. Many problems exist. First of all, it is more appropriate levels which are most relevant to our definition. In the instance with consciousness, one may define it according to macroscopic structures such as retina, eardrums, pain receptors. In a sense, one does not need the microscopic level, since there is not shown any atomic (or smaller) principle related to consciousness – perhaps some quantum theorists may disagree with me. Secondly, smaller levels require more precise measurements. If some minuscule error existed in our conception of microscopic laws, then any extrapolation to a macroscopic structure would be wrong. Third, how far at the bottom should we go? We all know atoms are not the smallest level of things. In sum, many macroscopic predictions have proven accurate without including any microscopic detail – we refer to cutting the heart in two pieces.

Artificial intelligence: some related reflections on free will, thoughts and consciousness

If we take the thermodynamic reasoning ad absurdum, it seems that people would behave very different when they consist of silicon, nanobots and other artificial materials (cyborgs), than when they consist of the human molecule – aside from the extreme increase in intelligence. I personally think that intelligence will be a larger factor in behavior, than is molecular composition. Why? Actions are determined by intelligence, which is determined by processor power and memories, which are determined by input, which is determined by external stimuli and sensory capacities. The stimuli outside of our bodies will remain (more or less) the same. The internal physical structure will differ somewhat, but if we have a memory unit with the same memories, a processor unit with the same capacities, and an action unit with the same behavioral capacities, then, disregarding molecular structure – behavior will be highly similar. Consider brain chips, prostheses, neural implants, pacemakers: although molecular composition changes, the actions will be the same – from a macro perspective, at least.

It may seem that this would imply free will. This only depends on definition. If free will negates physical determinism, it is clearly wrong. If free will means that physical laws dictate that some things contain information which can be decoded, and subsequently stored – and hence give rise to behavior that rests on this law, then free will exists. This is what is called *apparent free will*. Given some processor capabilities – physical structures that allow decoding of stimuli (decoding is of course nothing else than transforming some physical signal into another physical signal) – beings with apparent free will, have thoughts and can control their actions – albeit that their 'will' and their 'thoughts' are determined by the huge chain of events we call 'history'.

Does digital philosophy imply free will? If everything computes, what does that even mean? *The least* it means is that everything *can be modeled* as if it were a computer that manipulates symbols – for instance (0) when a property is absent, and (1) when it is present. In that case, whatever its underlying metaphysical truth may be, we know – at least if we can confirm this hypothesis- that we can predict anything with powerful cellular automata. Somewhat reminiscent of the 'extreme positivist' perspective, I think we should keep our interpretation to a minimum so the most we can say is everything can be modeled *as if* it were a cellular automaton.

Consciousness, defined as the integration of information, is a universal characteristic. However, one should not confuse consciousness with spirituality- and one should refrain from going to far on metaphysics. Even the measurement devices that are employed on Mars to analyze the soil, are intelligent, conscious, since they are able to analyze and store information. This ability derives from physical properties – chemical affinities that in no way have free will. This ability rather is a coincidental capacity of a molecular structure that gives certain signals when confronted with certain characteristics. Humans are very similar. We are programmed in a way to analyze things among other things (survival perhaps). Our analytic capacities are broader than those soil analyzers, however. In addition, both our hardware and our software are different. Different physical characteristics allow different speeds of computation (processing power) and different properties to analyze. At what point do 'thoughts' arise? Well, words seem to be important. However, even 'word' is a gradual contrast: birds can sing tones, which would equate with some meaning, for certain goals.

It is uncommon to bridge consciousness, thoughts, or information with physical determinism, since many assume that free will is real – perhaps because it is a quite complex analysis. Nevertheless, I think I have given some hints about how this *impression* of free will comes about: thoughts are construed by physical on-off switches, by symbol manipulating and property detecting mechanisms. At what point symbol manipulation occurs is somewhat unclear. In any event, non-biological things can have intelligence and consciousness as well – storing information, analyzing data, performing complex symbol-manipulating, building, or other tasks. A bridge between these concepts and physics entails nothing but changing one signal to another via chemical reactions.

Interesting applications of this reasoning

An interesting application of this reasoning in terms of physical chains causing our consciousness, is the study of sound waves as a means for mind control. It is shown that certain sound frequencies can used to put our mind into a certain state – and I believe Tesla was a pioneer in this. Another application for 'mind control' is of course manipulation of light waves, which enter our retina – as shown in the Hmolpedia. Other obvious examples include manipulation of weather, temperature. Of course, one may also consider upgrading our brain via brain chips etc.

Life and consciousness, a reflection in the light of immortality

Following the reasoning of the singularitarians, it seems necessary to embrace that 'being alive' often does not refer to a physical state, but rather a continuity of consciousness. Much of the atoms, cells and organs can be replaced while giving still the same sense of identity. When the singularity will occur, our biological organs will be gradually replaced by non-biological AI stuff. Nevertheless, we will still see ourselves as 'alive'. It seems our 'mind file' - that is, all the information we have inside of us available for use in the form of 'thoughts' - is key to consider ourselves alive. Suppose a part of our brain starts to dysfunction, because of an accident or some disease. We can insert neural implants to regain function, and we will feel 'alive' again. Suppose now that we gradually add non-biological parts to our body, which increase our function, and ultimately fully replacing the biological body. We will still consider ourselves alive – but are unlikely to call ourselves biological though.

Let us ask a different question: Would you mind consisting of different components, and nevertheless feeling the same (or better)? Conversely, would you find it pleasant to hear that we had a method of ensuring that none of your components get exchanged or upgraded, a downside being that you wouldn't be conscious anymore? Both answers are a clear no. If you disagree, try and disprove me by living up to your philosophy: freeze yourself in alive.

The debate again results in a Gestalt-analysis: very easily different components can give a same result ; and very easily same components can give a different result. Besides the obvious influence of differential environments, there is the relative configuration of components. I am, then, relatively sure that, if you have a rough human molecular formula, that one will not arrive at a human by synthesizing some huge molecule. The best part to start is from a few cells with DNA, as evidenced by pregnancy, cloning, stem cell research. In fact, I dare question the fact that such a huge synthesized molecule will remain *one*.

Apparently however, I would have to add another criticism, since Libb apparently goes further and further to remove words that he believes do not fit in 'modern science'.

The thinker who holds-fast to the ancient mythological doctrines of 'life', 'soul', 'consciousness', 'free will', 'choice', a 'brain', etc., will argue, to their grave, that, in some contrived-way or another, at one particular second in time, in the course of human evolution mechanism, that molecules, somehow, came to life, acquired souls, developed a free will, obtained the a state of consciousness, evolved the ability to think, among other now-defunct traits that do not apply to the hydrogen atom, nor to any other molecule, known in science.

Apparently, the brain is inexistent. Consciousness in some respect could be considered a marginal term, depending on its definition. But a brain? The ability to think? Apparently, Libb does not even consider the possibility that writing, thinking, planning exist, even if he is doing it by definition. Libb does not even consider that a 'human molecule' could be different from a hydrogen molecule.

To Libb: I have noticed you cutting yet another term: 'brain'. Why so? Suppose we damage your brain and inject you some molecules so that your average human molecular constitution remains the same. Will you behave the same way? Do you dare to experiment ? ;)

So what's next? A 'tree' doesn't exist, because it's not precisely defined in terms of molecular composition? A 'computer' isn't either. Following such a reasoning, we might as well conclude that ' city' or the 'earth' and yes, even the 'human' don't exist. Again, we refer to the notion of *effective definitions* and *Gestalt*. For effective definitions, we can define things according to an effect that this things has. An obvious instance is a motor: although not related to one specific chemical formula, it only distinguishes itself in terms of its effect (to propel something) , or perhaps also its shape. For Gestalt, we can see that a hydrogen atom and a human molecule behave very differently, since large combinations of elements can lead to surprising resulting behaviors.

The debate catches fire again

Libb: *Re (D.Boss): "To Libb: I have noticed you cutting yet another term: 'brain'. Who so?", I will refer you to the "mind from matter" section of the Charles Sherrington article:*

http://www.eoht.info/page/Charles+Sherrington

In other words, once you jettison the "life" concept, as Sherrington clearly did, in 1938, the next puzzle to tackle is the "mind/brain" issue. In Sherrington's own words:

"But if there be no essential difference between 'life' and all the rest, what becomes of the difference between mind and no-mind."

David:

[To Libb]mind perhaps is a different term, it has some different connotations. but once you say 'the brain does not exist', you can start throwing away all existing categories. medical science will come to a halt, because we cannot acknowledge that the brain exist. so on the one hand your proposal may be to jettison the term brain and use no replacement. another proposal you may make is to **again** use another term. in this case, inventing a new term is relatively useless, since the term brain has been shown to be more fruitful than any other approach to explaining the complexity of our behavior. cut just one area of it, and be numbed, or retarded for the rest of your life.

[To Jeff]I'm not sure what your reasoning is to debunk the fact that DNA is present in nearly all things that are considered 'alive'. also, the cell criterium of course, and perhaps some peripheral criteria you haven't added into the analysis.
I must of course agree that there is a lot of vagueness in many of the terms we use: life, consciousness, brain. Brain, I suppose is the best defined, but of course is not always chemically the same. Neither is the human molecule by the way: if the brain is chemically ill-defined, then the whole thing is even worse defined.
Part of this debate is at what level we consider things; Clearly at the level of organs, actions, etc. we can see with 100% accuracy when somebody is 'dead' - try testing this ;-). Or how about the level of 'thoughts'? One may dismiss 'thoughts' or 'intelligence' as being ill-defined, but clearly they exist - because there is something that makes us do these clever things - all resulting from our brain, no other organ is responsible for it.
Another important thing that may be the reason why I take such an ambivalent stance, is the famous Tesla quote: 'No thing is endowed with life'.
I actually agree with this, even considering my history of 'defense': life is not a fundamental property, it is an emergent property, after added hierarchic layers. Therefore no thing is *endowed* with life. Tesla goes on by saying something supporting determinism (the thing with the 'cosmic balance disturbed'). If this is what *life* is, then I am a non-believer. From there my ambivalent stances (life vs biology).

[To Libb] A thing that has deeply disappointed me in human thermodynamics is that I've come to think that using a minuscule level to extrapolate 20 levels up is bound to give errors. If we start from 'macro-certainties', such as 'my heart functions as a blood pump', we can derive certainties such as 'my blood won't flow if my heart stops'. Applying thermodynamics is useless, unnecessary and too difficult in this case. My thoughts have come more closer to Ernst Mayr, in that I think different objects of study require different methods. Humans can best be studied by history and neuroscience - at least if we want to make some useful predictions, what science is about. Some claim to have done this; Gottmann for instance made 'predictions' under the veil of thermodynamics, but in fact 1. he did not predict, he made a mathematical model *after* he had the results 2. the model had nothing to do with thermodynamics, he just said 'something like entropy exists in marriage' and such things - he used the 'number of positive events' or something, which is as vague as pseudoscience can be. Can you restore my confidence, Libb?

[To Jeff] Helium and the 'Human Molecule' (if this even is a correct name... ask yourself: if I place water into a bottle, does that mean we now have the bottled water molecule?) follow the same laws of physics, but that does not mean that they react the same to those laws to which they are subject.

Jeff: David, I agree that Ernst Mayr makes many good points regarding the difficulty in extrapolating some physics principles upward to the individual and ultimately, the ecosystem level. Another excellent macroscopic approach is Bejan's constructal theory, which has shown general validity across a wide swath of animate and inanimate processes.

To clarify my position, thermodynamics is the science of energy. Historically, it has had a focus on the study of processes involving the transduction of energy between heat and work. The scale of classical thermodynamics is macroscopic, however the focus of much contemporary study is on chemical thermodynamics, which tends to deal with systems smaller in size and with large number counts of interacting entities.

I am however, cautiously optimistic that we can use energetic principles in concert with new insights to develop a rigorous thermodynamic theory of ecosystems.

We are certainly not alone. There has indeed been much work already done. Peter Mauersberger's application of Planck's theory of dilute solutions to aquatic ecosystems was an important move in the right direction. However, we have plenty of work to do in bringing the often disparate sources of information together. For me, this is what Libb and the EOHT have already to a large extent achieved.

David: Very true, I can only encourage people to think in a multitude in directions, apply unique perspectives to gain a solid interdisciplinary knowledge frame work. This is one of the reasons the Hmolpedia is so cool.

Libb: Re: "A thing that has deeply disappointed me in human thermodynamics is that I've come to think that using a minuscule level to extrapolate 20 levels up is bound to give errors", this comment gives way to the view that you do not see the big picture (the forest amid the trees), and of course this is the case for anyone who has not yet had the opportunity to let thermodynamics soak into their mind for some time. Thermodynamics is a language built on physical axioms, of which there are no "scaling" factors involved. The foremost of these axioms is Boerhaave's law:

http://www.eoht.info/page/Boerhaave%27s+law

which means that heat causes bodies or volumetric regions to expand and cold actuates the reverse process and when this occurs energy is conserved and entropy increases. The entire structure of thermodynamics is built on this axiom. It makes no difference if the region is sub-atomic (e.g. quark-gluon thermodynamics) or super-galactic (e.g. black hole thermodynamics), the same rules apply to each. This is what is called the black box approach:

http://www.eoht.info/page/Black+box+approach

Also to give you some idea that you are way off the dart board in your ideas about thermodynamics, the entirety of the subject is derived from the study of the operation of the heat engine, the visually-grandest of these being the 1678 gunpowder engine, which as you plainly see:

http://www.eoht.info/page/Gunpowder+engine

is some 3 or 4 bigger than human-level, which means, loosely, that we are extrapolating down from the macro to the micro, rather than up as you seem to understand the historical derivation of thermodynamics. The chemical thermodynamic formulation of thermodynamics is relatively recent, a subject that began in the 1870s.

The takeaway point that you will want to hold on to, from the previous post, is that just as sodium (Na) will "spontaneously" react explosively with water (H20), whereas sodium chloride (NaCl) will not, so to will 20% of people "spontaneously" fall in love at first sight and marry that person:

http://www.eoht.info/page/Love+at+first+sight

whereas 80% will not. The measure of the "spontaneity" in each case is Gibbs free energy. This can be explained in terms of chemical affinities, as Goethe did in 1809, or in terms of free energies, as we are doing in modern times:

http://www.eoht.info/page/Human+free+energy

or in terms of both affinities and free energies as Walther Nernst did in 1893:

"Since every chemical process, like every process of nature, can only advance without the introduction of external energy only in the sense in which it can perform work; and since also for a measure of the chemical affinity, we must presuppose the absolute condition, that every process must complete itself in the sense of the affinity—on this basis we me may without suspicion regard the maximal external work of a chemical process (i.e. the change of free energy), as the measure of affinity. Therefore the clearly defined problem of thermo-chemistry is to measure the amounts of the changes of free energy associated with chemical processes, with the greatest accuracy possible ... when this problem shall be solved, then it will be possible to predict whether or not a reaction can complete itself under the respective conditions. All reactions advance only in the sense of a diminution of free energy, i.e. only in the sense of the affinity."

http://www.eoht.info/page/Walther+Nernst

Spend some time reading Lewis' thermodynamics textbook to convince yourself of this.

To repeat Nernst, in respect to human chemical reactions:

http://www.eoht.info/page/Human+chemical+reaction+theory

"Therefore the clearly defined problem of [human] thermo-chemistry is to measure the amounts of the changes of [Gibbs] free energy associated with [human] chemical processes, with the greatest accuracy possible ... when this problem shall be solved [in the coming centuries], then it will be possible to predict whether or not a [human chemical] reaction can complete itself under the respective conditions."

It cannot be stated any clearer than this. In short, someone in the coming millennium, will figure out how to accurately measure changes in free energies associated with human chemical reactions, whether in love or war, and in the great words of Nernst "this problem shall be solved". That day will be an excellent moment for whomever is fated with that destiny.

David: [what I mean with extrapolate up 20 levels:] for instance, the human molecular orbital hypothesis assumes that you can extrapolate quantum physics up to the level of macrostructures

such as humans. it even assumes that this human molecular orbital overlap is the cause of 'moving in together' or 'getting married'. this is what I would mean with extrapolations of 20 levels up: first you use a quantum physical concept, extrapolate it up to the macroscale, and consequently apply it to a cognitive-social-metaphysical concept.
a second instance is where you think that humans 'bond' similarly like atoms to each other. Clearly, humans do not bond just as atoms or molecules- otherwise couples wouldn't be able to go to work separately. So the depiction A+B-> AB is somewhat untrue, even assuming the human is a molecule.
a third instance is that the molecule requires that atomic bonds are established. if we pour water into a bottle, it does not mean we have the bottle+water molecule... similar goes with humans.

I think it would be much more interesting to study memory and other brain processes to find out what 'love', 'hostility' and such really represent in terms of physical happenings. our brain, after all, is the tool that, from our perceptions, creates these emotions.

Maybe you will find something interesting. let us hope.

An interesting comment is that Libb seems to have changed his mind. In earlier discussions he emphasized that the bottom-up way is less biased. Indeed, there is no reason why bottom-up would be preferable if there is no difference in predictive power. This predictive power is maximal in some cases in medical science or neuroscience (take out X area of the brain or Y area of your body, and Z will happen) and, or even in behavioral science (at 9 o' clock I go to work, for instance).

The issue that Libb will not be able to resolve, without acknowledging the existence of 'brain', is that he cannot explain the most simplest actions in daily life: the mere act of reading you have an appointment will cause you to go to it; not some macroscopic thermodynamical attraction between the 'building molecule' and 'my human molecule'. Mere cognition is enough to predict behavior. Of course, physics may be used to explain what precisely happens in the brain. This is an issue that is much more likely to be of importance than a 'human molecular bond'. On the human bond via human molecular orbitals, I return in the last chapter.

Determinism and our brain

When people talk about determinism, it does *not* mean that we have no brain or choice – and thus that a psychological analysis, an analysis of thoughts would be useful. It means that our brains and thus our choices are determined by a long chain of events. Yes, many of those causal events can be described by thermodynamics, that is not the point. The point is that the brain can be described in other – more useful in terms of predictions – methods. Freud has tried to make something out of the thermodynamics-brain connection but has little to show for. Neuroscience and AI on the other hand have helped people recover from neurological disease, etc. But I bet neurological disease is a defunct concept, according to Libb? Apparently Libb does not care for predictive or useful science, and prefers to call every term that is not directly related to chemical thermodynamics 'mythological' – an easy way out of an argument that is not related to his favorite topic.

Yes, we can make accurate predictions of human behavior without using any thermodynamics. For instance, having the knowledge that a certain person has a 9 to 5 job at a certain place X and that lunch breaks are held at work, we know with high probability (say 99%) that the person will be in that place X on Monday at a random moment between 9 to 5. Similarly, other extrapolations can be made using nothing else than this historical method. In fact, I have yet to witness an accurate thermodynamical prediction of human behavior – yes, Gottmann's study was flawed : it didn't *predict* anything, but rather post hoc made some equation to fit with the data. Neither did it rely on thermodynamics. In this study, thermodynamics was used as a metaphor, saying "Something like a second law of thermodynamics seems to function in marriage—that is, when marital distress exists, things usually deteriorate (entropy increases)."

A conclusion: No, we don't need entropy to explain the effects of stimulation of some neural area. Yes, our brains are most important to understand what we will do next.

What? Our brains are not well-defined? Even if one adheres to the weird vision that for something to be well-defined it must be defined in terms of chemistry, it is clear that a 'brain-molecule' would be *better* defined as a 'human molecule', since it is a part of it. Furthermore we can insert devices into our body, such as pacemakers, which are not in our heart – indeed the human molecule is badly defined.

This is why this one-sided approach does not work – there would be no definitions, aside from 'here we have a clump of CHO2HO ...' with anything we encounter. Trees do not exist, they are mythological constructs !

I might imagine a definition of the brain by Libb : a hydrocarbon molecule. The problem with this reasoning is that this is in no way a feature that distinguishes the brain from other hydrocarbon molecules. To define something is to define its most distinguishing characteristics: we shouldn't say a lamp is defined by this or that molecule – it doesn't have one single 'lamp molecule'; all lamps are different; what defines a lamp is its 1. Manmade nature 2. Functions to provide light 3. Is able to be switched off and on; Similarly, what is unique to a human being is not its gross molecular composition: it is its biological ordering of cells and organs, and the arising macro-behaviors that may be called intelligent. So the approach gets us nowhere; Clearly, one can formulate a human molecular formula. This does not mean that any clump with the same elements – that one would not consider human - will behave the same. Furthermore, such a formula denies the fact that molecules are defined as bonded atoms. One can imagine that the water in our blood for instance is not atomically bonded with the hormones, sugar, carbon dioxide etc. Rather, grains of solids are mingled with grains of fluid, with no bond being established. The water itself that is in me is not bonded by electrostatic forces as is necessary to form atomic bonds, but rather, it resides in my body – as is evidenced by blood flow if one cuts oneself, and by urine flow if our bladder is full. We might as well put two bottles in vicinity, with the one bottle inside the other, and consequently say: 'voilà, the two-bottle molecule.' Clearly there is no electrostatic force that bonds the bottles. The only factor is vicinity.

Libb's unique way of seeing free will is also a very flawed one: according to Libb, our brain is not crucial to our behavior, our brain does not decide what we do. This is *not* what the free will debate is about. The debate is not about whether our brain is central executive of our body. Clearly it is. The debate is whether our thoughts, the computations of are brain, are fully determined by past causal events, or whether our behavior can be predicted should one know all the laws of nature and the initial conditions of a system. The answer to this is clearly that free will does not exist, since everything follows physical laws.

Free will

In the "Automaton theory", Descartes posed that animals are 'automated'. Unfortunately, in his dualistic vision of man's intelligence, he posed that 'mind' somehow detaches from the material world to control the flesh of man. We should cut the anthropocentric part of the latter.

Free will exists not at the fundamental level, but at the *apparent* level. If I want to move my finger, I can do so. That is not the point. The point is that, given the laws of nature, I, in theory can be fully predicted by a hyperintelligent being.

We may regard the present state of the universe as the effect of its past and the cause of its future. An intellect which at a certain moment would know all forces that set nature in motion, and all positions of all items of which nature is composed, if this intellect were also vast enough to submit these data to analysis, it would embrace in a single formula the movements of the greatest bodies of the universe and those of the tiniest atom; for such an intellect nothing would be uncertain and the future just like the past would be present before its eyes.

—Pierre Simon Laplace, *A Philosophical Essay on Probabilities*

So basically Laplace imagined a demon who knew every law of physics. This hyperintelligent being would be able to predict so many laws of physics, that he, at any given moment, could decide what the future would bring, and what happened in the past to cause what is happening right now.

The only way I believe 'free will' to exist is, for instance, with humans: their brain allows more wide range of behaviors. This is not true free will of course, but it can explain why the term free will seems intuitively correct. But free will thus only exists at the *apparent* level.

Determinism equals scientism

Suppose we would adhere to a philosophy of 'free will'. In this case, beings would be unpredictable, even in theory. Then science would have no way in predicting 'living' systems. So biology, zoology, anthropology, and psychology are essential useless sciences. There can, in this case, be nothing that predicts intelligent beings, for they have free will and they can do unexpected things.

But this is not what we see in the laboratory: if we stimulate the brain of an ape, it will do exactly as we predict (given that we know what the effect of stimulation is). So, the ape, instead of chosing every move he makes, is subject to the electrodynamics in his brain. Results are similar with humans.

Suppose now we adhere to a philosophy of 'determinism'. In this case, every being is, in theory, predictable. This seems

1. More in line with the previous paragraph.

2. A step closer in doing good science.

Let's evaluate 2. If we go of the assumption that, indeed, every being is in theory predictable, then the 'life sciences' actually have a meaning in science: it is possible that, one day, we will be able to predict the behavior of living beings. Since that is what science aims to do – make predictions- it should be clear that adhering to a determinist standpoint is the only way science makes sense: if atoms, molecules, objects, or beings had free will, then there is no use to try to predict them. Since we *are* able to predict the behavior of atoms, molecules and things, and , in theory, even the behavior of beings, it should be clear that atoms, objects and beings have no free will. If you do tend to argue that highly intelligent beings have free will, while lower species have not, then either a) there is a confusion in the definition in either one of us, or b) there is no confusion in the definition and you believe in two opposing principles at the same time.

Case a: Free will can of course also be defined as: because of human beings being so knowledgeable, they can select situations and opportunities more careful, and they can manipulate the environment more easily to their own wishes. In this way, humans have more free will than apes. However, in this context, we are having the debate of free will vs determinism, a debate about "are we dictated by atoms, chemical affinities, the forces of nature, and fixed underlying principles, rather than a true 'choice' where we are not determined by principles outside of us, but we choose with our 'free will' (some believe it to be residing in the brain, other even claim a soul)?"

Clearly I believe the first, since our brain is clearly a predictable structure with fixed underlying rules.

Case b: If you argue that only highly intelligent beings have free will, given the just given definition, then you are being inconsistent: you believe there to be fixed underlying principles for lower beings, but you believe humans to be 'special', an extremely anthropocentric vision. You possibly then believe that humans have a 'soul' or something (Descartes made this mistake I believe, he thought the rest of the animals were 'automata' (with this I would agree), but that humans have a soul, which sets them apart from the animals (don't agree)).

There is no reason to believe there are no fixed principles in humans. Even our brain is a very predictable organ: stimulate this region and you will do this. The human brain and the resulting behaviors are more subject to electrodynamics and the neural structure than they are to free will.

Perhaps Schopenhauer made a nice summary to the role of free will by saying we can choose what we do, but we cannot choose what we want.

On quantum physics and free will

The act of observation seems to collapse the wave function in so-called double slit experiments. Therefore, subatomic particles are particles when they are observed, and they are waves of infinite possibilities when they are not observed. It is as if the subatomic level is conscious. As if consciousness has power over matter. As if we create reality. At least according a select few quantum physicists.

When asked 'free will or determinism', quantum physics dictates the answer: free will. There is uncertainty. We choose, and our choice makes the wave function collapse. At least this is what Michio Kaku makes us believe.

Is this all a valid conclusion? Perhaps not. Of course low-level consciousness (as in 'reacting') which even ants could be explained by introducing the idea that quanta have some kind of 'awareness', and that they react to an observer '. Or ... you could explain this by the more simple postulate that chemical molecules react to each others presence and that this explains why even non-living things can react and seem to be conscious. How to explain the reacting to the observer, or rather the act of measurement, is not clear. Anyhow, it seems to me that quanta do not have eyes, and therefore are not 'aware' in the traditional sense that we are observing them. They might react as they do, but this has nothing to do with their 'awareness', no more than a molecule would react if another molecule is placed in its vicinity. If it should be the case that quanta have awareness, then these quanta have sensors. Clearly quanta do not have sensors. Their reaction to measurement should be explained in terms of cause (the measurement)-effect(wave becomes a single particle). Even if a quantum system is a very chaotic, complex system, of which it cannot be predicted what possibility (particle) will be chosen from the range of possibilities (wave), this does not mean that humans have free will- as in undetermined by their milieu. If you really wanted to predict a quantum system, you would have to go of the assumption that we know every possible interaction of this particle. We do not, and can therefore, not yet at least, predict the quantum system. That doesn't mean it is *truly* unpredictable. It is unpredictable to an imperfect being. The laws of nature follow rules, and so do we. We cannot choose who we love. We cannot choose what we do at this instant. Unless we literally think about doing something unpredictable, in which case we still would do the predictable (in this case a seemingly unpredictable behavior). All has been written in the laws of physics and chemistry primary, and secondary effects from the laws of habit.

It is a hard theory to ever test: the theory that quanta have free will. If they have, then we can only be sure (?) if their behavior is proven to be truly random (equal probabilities for each possibility). This is of course hard to prove 99.9999999999 % without extreme computational power. But, in reality, it should be clear that the world is not entirely random. So neither are quanta. But this does not disprove that quanta have free will. And, in retrospect, neither does randomness: if dice are thrown, does that mean they have free will? Not at all It means there exists a law, based on symmetry, for it being random: if you would create skewed, or unfair dice, the numbers would not have equal probability. So, even randomness can be explained by fundamental laws where there is no speaking of a 'force that chooses'.

I will now perhaps give a great answer into why all this mess of free will began in the first place. Answer: religious influence of invoking 'spirituality', a 'soul', a 'life force' that has a 'mind outside of matter', in an immaterial world. The whole debate on free will rests on the idea that there is this 'mind force' that is 'above matter', that can detach itself from its connection to the material world, resulting in a truly autonomous choice. Otherwise, the debate would have no meaning. Why should we even question the fact that everything is guided by fundamental laws? Are we not determined by the laws of physics and the power of our brains, our molecular affinities, and our genetical structure? Is, our genetical structure, in turn, not determined by the whole of history. Did our genes 'choose' to exist? No. Did we choose to exist? No. Did we suddenly spring to life, as an accident of billions of years of chaos and evolution? Yes. Even randomness can be explained by fully determined processes underlying it. Even randomness does not mean – notice the huge leap that one usually makes

– that 'choosing forces' exist to do what they 'want'. That's what free will implies: it implies that quanta have desires, as in a *truly autonomous* immaterial mind, to be in one place or in the other. Well, that might be the case, but hey, I would go with the more parsimonious explanation that quanta have no such desires, and that immateriality does not exist. I conclude that notions as the soul, notions as the spirit and life force are still very much influencing, very negatively, our thinking.

The only way I think free will exists in humans in relative sense to other beings if we define as follows : " allowing complex behaviors (read: a wide range of behavioral options)". These options are not transformed into choices, but rather into resultants of a intern and extern milieu that determined which option was most plausible.

One problem that may arise with Laplace's demon is the following. In the case of the dice, it is clear that there are inherent rules, and that, if you would calculate the position of my hand in comparison to the table, and the strength of my throw, and the shape of the dice, a hyperintelligent being could predict this process. Yet, this is not a truly random event. But the only evidence we have, as of yet, of truly random processes is not in nature. Perhaps it is possible in some computer simulations. But anything that is described in the true world to be random, is, in reality not: if we are playing cards, a hyperintelligent being that knew of all preceding actions would know exactly what cards will be dealt. The faster the shuffling of the cards, the harder to predict, and the more complex brains or other 'thinking stuff' (such as silicons) necessary. But in the simulated world, it becomes clear that cards are dealt by a random generator, and thus the process is random However, we have come to this randomness by a deterministic rule 'create a random number with spade, club, diamond, or clubs'. However the task is difficult to predict the outcome. But I have some idea of how this would happen. A being with sufficient insight, would know the underlying algorithms of the random card generator. In this case, it might be still predictable. An illustrations of how random generators in fact *are* predictable, is the following statement by 'random.org':

> Perhaps you have wondered how predictable machines like computers can generate randomness. In reality, most random numbers used in computer programs are pseudo-random, which means they are generated in a predictable fashion using a mathematical formula. This is fine for many purposes, but it may not be random in the way you expect if you're used to dice rolls and lottery drawings.
>
> RANDOM.ORG offers true random numbers to anyone on the Internet. The randomness comes from atmospheric noise, which for many purposes is better than the pseudo-random number algorithms typically used in computer programs.

So still, a random generator that is called 'truly random' derives from athmospheric noise. This might make us think that athmospheric noise is a truly random process in nature. If this were true of course, Laplace's demon would be falsified. However, I believe it to be a big mistake to say that something is truly random, just because we did not yet find any algorithm to generate it. Human foolishness is sufficient enough to explain why we still think such a phenomenon to be random, and also why we haven't found the underlying pattern.

In any way, whether Laplace's demon is true or not, that is, whether an omniscient being would truly be able to predict everything (I'm guessing yes, since omniscient, after all, means 'all knowing'), this is not even the issue. What it does mean is that there is no 'choosing life force' which somehow detaches itself from the material world, so that its choice is truly autonomous. Perhaps many people do not like this definition of free will, because they know this is impossible, and they like to believe they are truly free. Perhaps some other people do not like my mocking of the soul, because they like to believe in fairy tales like the afterlife, that are, let's be honest, are wishful thinking with no empirical evidence going for it.

The correct formulation of the uncertainty principle is also important. "If you *perfectly* know the amplitude distribution on position, you *necessarily know* the evolution of any blobs of position over time" The name 'uncertainty principle' gives the impression that we cannot know the position of the particle. However, this uncertainty comes only because of imperfect knowledge.

I have made the reasoning in this paragraph that things that seem random are only pseudo-random. Jeff Tuhtan though has made an interesting reverse speculation: how can we distinguish seemingly deterministic patterns that are truly random in nature (pseudo-determinism), from truly deterministic patterns?

That may be indeed the case as well- such as when you flip coins that are unbiased for a small number of times, and there is an accidental pattern in this particular dataset. Similar it might be for many rules we find - that they are temporary pseudo-rules that actually stem from a random process. However, as many basic laws that we are certain of keep holding true for quite a while now, those are certainly not random. Because of the 'science works' criterion, we can be certain of some deterministic rules. Can we be sure of randomness? I think not. How on earth are you going to prove that there is no rule for it? You would have to make an inventory of all the possible rules and check them all, which is just impossible.
Also, I think it is bad for science to assume randomness, because it equates with giving up to find a rule.

Predicting human behavior

One problem with any paradigm trying to predict human behavior is the fact that we cannot fully predict systems that are equally or more intelligent than we are. Not only are humans not smart enough to know what all their peers' motives are, but also, once we might establish a comprehensive rule, it might be the fact that these peers adapt and refuse to follow the rule we have discovered. But even without all these, it is still easy to see that human thermodynamics disregards any cognitive factors. Although the philosophy that everything is chemical is of course undeniable, people are determined by their previous relations and the content, potential and inclinations of their own neural networks and capabilities. Behavior is too complex to predict via just a few equations.

Yet it is shown, by the principle of cellular automata, that, with a few simple rules, there can arise very complex behaviors over time. But, to make any precise prediction, in this case, you would have to be able to replicate every given memory and capacity of each individual to predict the behavior of this human. In this paper, it is argued that sufficiently complex cellular automata might aid in the process of actually *precisely* predicting human behavior. However, the computational capacities of computers are too low, for now. Another challenge is that we have to provide the correct initial conditions: our science must be absolutely flawless. If any flaw in our theoretical assumptions exists, the end result will be very different.

Quantum physics: random behavior?

The vision of free will relies on ages of spiritualism and dualism. Descartes was one of the philosophers to introduce the concept of 'res cogitans' (the thinking thing) which basically was the same as the notion of the soul. This res cogitans was uniquely human and was based in an immaterial world, and only via the epiphysis, could it interact with 'res extensa', the physical body. Animals, Descartes argued from a very anthropocentric standpoint, are different than humans, in that they have no res cogitans. Therefore animals were seen as automated machines, subject to the material world, whereas humans were somehow different. It should be clear that, if we substitute res cogitance with life force, or free will, or soul, that we have the same result. This result is inconsistent: it assumes some beings have contact with an immaterial world, whereas others have not, a somewhat egocentric, but also unfalsifiable and unscientific standpoint which will never have empirical evidence going for it.

Another such attempt to invoke free will involves modern quantum physicists. The general idea is called "consciousness causes collapse". In the 1960's, Eugene Wigner reformulated the "Schrödinger's cat" thought experiment as "Wigner's friend" and proposed that the consciousness of an observer is the demarcation line which precipitates collapse of the wave function, independent of any realist interpretation. Wigner identified the non-linear probabilistic projection transformation which occurs during measurement with the selection of a definite state by a mind from the different possibilities which it could have in a quantum mechanical superposition. Thus, the non-physical mind is postulated to be the only true measurement apparatus. This interpretation has been summarized thus:

"The rules of quantum mechanics are correct but there is only one system which may be treated with quantum mechanics, namely the entire material world. There exist external observers which cannot be treated within quantum mechanics, namely human (and perhaps animal) minds, which perform measurements on the brain causing wave function collapse."

However popular the vision in the layman community might have become, most quantum physicists disagree with this interpretation and even Wigner- who has changed his stance - has admitted that he has faultily applied microscopic scales to macroscopic scales. But as should be clear, this vision that Wigner has produced is, in effect, dualism. It introduces the concept that mind is fundamentally different from matter.

Other attempts to reinstate free will are the attempts to prove that some processes are random. It should be clear that very few processes are random, otherwise we won't have so much successful predictions in science. In this paper, it is proposed that 'determinism equals scientism', meaning that, following a non-deterministic philosophy, one can only conclude that science, which aim is to predict, is useless. Since predicting is only possible if the world follows deterministic rules, science is useless in such a world.

Anything that is described in the true world to be random with regard to some property, is, in reality not: if we are playing cards, a hyperintelligent being that knew of all preceding actions would know exactly what cards will be dealt. The faster the shuffling of the cards, the harder to predict, and the more complex brains or other 'thinking stuff' (such as silicons) necessary. But in the simulated world, it becomes clear that cards are dealt by a random generator, and thus the process is more random. However, we have come to this randomness by a deterministic rule 'create a random number with spade, club, diamond, or clubs'. However the task is difficult to predict the outcome. But I have some idea of how this would happen. A being with sufficient insight, would know the underlying algorithms of the random card generator. In this case, it might be still predictable, if it is not truly random. An illustration of how random generators in fact *are* predictable, is the following statement by 'random.org' (my bold).

*Perhaps you have wondered how predictable machines like computers can generate randomness. In reality, most random numbers used in computer programs are pseudo-random, which means they are generated in a predictable fashion **using a mathematical formula**. This is fine for many purposes, but it may not be random in the way you expect if you're used to dice rolls and lottery drawings.*

The same website then claims to offer truly random numbers.

RANDOM.ORG offers true random numbers to anyone on the Internet. The randomness comes from atmospheric noise, which for many purposes is better than the pseudo-random number algorithms typically used in computer programs.

This might make us think that athmospheric noise is a truly random process in nature. If this were true of course, Laplace's demon would be falsified. However, it is quite conceivable, that we cannot differentiate between 50% and 50.000001 % - in the case there are only 2 options. Underneath seemingly random processes can lie a very deterministic rule- as will be shown in the next chapter on complexity theory- which the maker or knower of the rule would know. A sufficiently complex rule is inseparable from randomness to most observers.

If there are *truly* random processes – where there is no rule, however complex, underlying - in nature, it is my guess that one cannot predict every move I will make, thus falsifying Laplace's demon. This does not mean that we make fully autonomous choices, free from influence of the material world – what free will would imply. This would only mean that the process specified is somehow detached from predictable influence. Suppose athmospheric noise is *truly* random, then does this change the predictability of events that do not relate to athmospheric noise? It does not seem so, since science has a long history of successful predictions.

A strange view presented by some quantum physicists is the many-worlds interpretation of quantum mechanics, which asserts the objective reality of the universal wavefunction, but denies the actuality of wave function collapse. Many-worlds implies that all possible alternative histories and futures are real, each representing an actual "world" (or "universe"). So, in effect, following this interpretation, it is argued that everything that can happen, *does* happen. However, it is an untestable concept, so we cannot prove it. The same goes for the Copenhagen interpretation where the act of measurement causes the collapse of the wave function. Both interpretations are philosophical arguments we cannot really test.

Something we *can* test, is whether our scientific predictions work. Apparently, our world is very predictable, given all scientific progression. A deterministic world implies that cause-effect chains are predictable, and this is exactly what science achieves. Thus it seems necessary to explain randomness for what it is: *apparent randomness*. Any rule that is complex enough, multi-factorial, is indistinguishable from randomness. That does not mean it *is* random !

Determinism : an indication by complexity theory

One approach to answer the question of determinism and randomness is chaos theory. Chaos theory studies the behavior of dynamical systems that are highly sensitive to initial conditions, an effect which is popularly referred to as the butterfly effect. Small differences in initial conditions yield widely diverging outcomes for chaotic systems, rendering long-term prediction impossible in general.This happens even though these systems are deterministic, meaning that their future behavior is fully determined by their initial conditions, with no random elements involved. In other words, the deterministic nature of these systems does not make them predictable. Indeed, rules *can* underlie seeming randomness, as was proposed above.

It has been shown that algorithms are a broader - or at least, an easier to handle - class than are mathematical rules, since there are actions that can be done algorithmically, but not mathematically at this moment. Computer science holds great promise for predicting all sorts of phenomena.

Is quantum theory wrong?

Before making dualistic speculations, that mind does not follow from matter, quantum physicists should wonder if their theory is really complete. I believe it is not. It is not wrong either. It is incomplete. Still, some examples might show that the foundations of quantum physics might begin to shake and shudder.

Randell Mills, claims to have built a prototype power source that generates up to 1,000 times more heat than conventional fuel. Independent scientists claim to have verified the experiments and Dr Mills says that his company, Blacklight Power, has tens of millions of dollars in investment lined up to bring the idea to market. And he claims to be just months away from unveiling his creation.

The problem is that according to the rules of quantum mechanics, the physics that governs the behavior of atoms, the idea is theoretically impossible. What has much of the physics world up in arms is Dr Mills's claim that he has produced a new form of hydrogen, the simplest of all the atoms, with just a single proton circled by one electron. In his "hydrino", the electron sits a little closer to the proton than normal, and the formation of the new atoms from traditional hydrogen releases huge amounts of energy.

This is scientific heresy. According to quantum mechanics, electrons can only exist in an atom in strictly defined orbits, and the shortest distance allowed between the proton and electron in hydrogen is fixed. The two particles are simply not allowed to get any closer.

(source: http://www.guardian.co.uk/environment/2005/nov/04/energy.science)

In any event, before jumping to all kinds of conclusions, we must know that our knowledge of the universe is incomplete. To postulate 'indeterminism', which equates with 'not fully determined by natural laws', is somewhat mythological – and it is in my opinion the last thing we should consider.

Thoughts: analyzed computationally and metaphysically

Something however, where I think Libb has problems with is 'thoughts'. Clearly an anthropocentric term, Libb might have some problems with this. Nevertheless, we cannot deny that there the word 'thoughts' refer to something that exists, even if it were different than what most would imply- hence on such an occasion, I would keep the term thoughts and equate it with some scientific definition.

Libb would probably think of thermodynamics in his definition. I would rather use computational models, since that is more relevant to cognitive science. However, we must consider an integration of methods, in an interdisciplinary fashion.

The mind is something strange, really. The mind allows a certain fraction of control. We cannot control what is in our mind, but our mind can control actions, and, in a futuristic perspective, can even control things without touching them (telekinesis).

Metaphysical things like thoughts and emotions exist physically in the sense that they are caused by physical structures. However, the phenomenological result is not present in the physical world, since 'emotions' and 'thoughts' are internal representations of things that are not really there. These things do correspond to physical happenings, but what the phenomenology of these happenings bring is entirely different in terms of qualities of what actually happens. There is for instance no sound and still our brain can perceive this to be there. Furthermore, emotions and thoughts do not really occur in the visible, touchable or hearable world, and that's why metaphysical applies. Well, approximately… : we *can* sense emotions and thoughts, but we do not know via what sense.

A best way to describe this sense, however, is as the 'brain sensor' – which reflects that some sensors in the brain transform or integrate information of previous events into active representations. Neurons fire impulses all over our brain, while nevertheless we 'see' things, or 'hear' things that are not there. 'Seeing' and 'hearing' are merely transformation of sensory input of 'light waves' and 'sound waves' respectively. And of course, these internal representations that are composed of these experiences of sound and light function by the same neurons – and other ones as well. Emotions and thoughts however, are more than just a repeated sensation. They are formations of abstractions that are compositions of the previous sensory inputs. Thoughts can be considered as integrated sensations (or better, representations in the now) based on the now and the previous memories, where sensation is the transformation of sensory input, and sensory input is any internal or external stimulus that can excite neurons in one particular way – depending on which 'sense' we are talking about...

Where the term emotion and thoughts get really metaphysical, is that they are both abstractions of the thing beyond these physical actions. They refer to what is presented in the mind: the brain has formed all sorts of associations based on the previous memories of sensations. It has thus constructed a visual-phonetic representation that never really existed in reality. It is not what is occurring in the brain that is thought to be metaphysical, it is what we perceive *because of this* that is metaphysical: we perceive this visual-phonetic representation of something that did not happen in reality. Even if it did happen in reality, it did not happen at this particular moment. Surely, a neural network underlies this representation, but this representation itself is hard – though not necessarily impossible – to anchor in terms of physical principles. For now, we thus call this representation 'metaphysical'. Hopefully, we will one day be able to know this 'principle of thought'. In fact, I believe Kevin Warwick has come close to this, by allowing sensations of the brain (moving-your-finger sensation) to transfer from one nervous system (his wife's) to the other (his own). Most likely the principle of thought is based on simple laws we already know, such as electromagnetism and light and sound waves.

Metaphysics and Plato's idea world

Several metaphysical things exist, but not in the way we make them to be. For instance, we have seen the example of *thoughts*, which, in the physical world merely are neural patterns. In the world of phenomenology however, thoughts in the sense we usually mean – a dreamed reality – do exist. This dream world is generated by the neural patterns but it is unclear how it arises from them. Not to say that there is a mind-body problem. It is clear that without these neural patterns we would not have those thoughts, but it is also clear that 'thoughts' is an abstraction of a dream world, where in reality there is merely a firing of neurons happening, which not incidentally, relate to previous happenings that caused similar sensations – we have seen how thoughts are integrated patterns of previous sensations. This is also not to say that 'immateriality exist'. Thoughts as perceived in phenomenology do not exist in the real world. It is this category that is probably inexistent in the physical world, just as 'moral behavior' is inexistent in the physical world. Nevertheless, both can be – albeit roughly – correlated with neural pathways and behavioral reactions.

Why do we *make* these 'inexistent' categories? We invent the abstract prototype 'morality' to avoid suffering, which can be defined as activity of the pain receptors – even mental 'pain' truly activates these pain receptors. This is the number one moral consideration: if you don't want to avoid suffering of others, you are called a psychopath.

Another such abstract prototype is 'friendship', or 'good'. While there are deeds that are typically considered to good, or, in line with friendship, these terms are human made abstractions which have no clear physical correlate.

In the case of friendship, love, or good, there are some of its connotations however that may be physically existent. I can see how biochemical reactions as well as light waves and sound waves, are related to good friendships and good love.

Mathematics is perhaps an even better example of the abstract prototype which does not really exist in the physical world – before we decided to invent it. The abstraction of the ideal mathematical path along which nature works, in the real world, has some variance or error to it. This is what Plato wanted to say to us, I think.

What if these things are not made – but actually represent, as Plato envisioned, an idea world of abstract prototypes, of which the happenings in the world are merely shadows. It is as if the metaphysical world is the world of perfect mathematical rules, whereas the true world also contains errors in these mathematical rules – call it noise, variance. Still, that does not mean this metaphysical world is reality. It is the abstract rule behind reality.

To clarify my point as to how metaphysical things exist and at the same time do not: dreams are real in the sense that they result from real processes, and in the sense that the dream itself is a real process. However, this representation of the dream – the idea that you are for instance playing soccer, while you are in deep sleep – is not according to reality.

This of course goes back on the debate of how our sensory mechanisms relate to reality. How do we know that what we perceive is reality? There are estimates, says Kevin Warwick, that we only sense about 10 percent of what is really going on. Furthermore, Descartes has mentioned the possibility of a 'malin genie', who could be deceiving our senses.

Not again? The life debate?

There is one quote from 'The Singularity is Near' by Ray Kurzweil that I think is relevant to this debate. In this section, we will adopt a different approach to the life debate.

'Robert Freitas estimates that eliminating a specific list comprising 50 percent of medically preventable conditions would extend human life expectancy to over 150 years. By preventing 90 percent of medical problems, life expectancy grows to over five hundred years. At 99 percent, we'd be over one thousand years. We can expect that the full realization of the biotechnology and nanotechnology revolutions will enable us to eliminate virtually all medical causes of death. As we move toward a nonbiological existence, we will gain the means of "backing ourselves up" (storing the key patterns underlying our knowledge, skills and personality), thereby eliminating most causes of death as we know it.'

I think we will all can intuitively understand why we will live over one thousand years, when 99 percent of diseases related to aging are resolved. However, the non-biological existence is somewhat more controversial. Although there are many post-humanists who swear that an uploaded mind is still 'you', there is really a problem with the definition of many of the terms we use.

First, we have the definition 'you'. What am I? Well, in terms of physical properties, it is clear that an uploaded mind is not me, since most of my biological body parts are replaced by totally different non-biological materials. It's also clear, however, that my biological body is not an unchanging piece of material. After each passing second, your body will be of different composition than it was before. It seems, from human's desire for their thoughts to be preserved, that we see 'Me' as *continuity of our thought patterns*. So, this definition is hard to anchor in a physical definition. I think we can rightfully label this as a 'metaphysical' definition, since 'thoughts' are an abstraction of a process that is difficult to pin down physically, but we all know it exists, however ill-defined - compare with the term 'love', 'friendship', and 'death' (see below). These thoughts operate via physical principles of course, even if those principles are hard to pin down. What else is somewhat 'metaphysical' about the term 'thoughts', perhaps even more so than the previous, is that there exist sounds and images without these actually appearing in physical form. It is clear, to me at least, that much of our morality is based on such metaphysical notions such as 'honesty', 'love', 'friendship', and so on.

Similar, apparently, goes for our notion of life. We don't care whether we consist of the same atoms as before. What does matter, is that we have the same sense of identity (which, again, is a metaphysical construct). Our bodies seem not to matter that much; what is important is that our brain is preserved ; or perhaps more accurate, that our thoughts - which form our identity, our pride and our experience - are preserved. Our brains are the most complex organs in a radius of at least a few light-years. If we can improve this organ by means of non-biological tools, and keep our sense of identity, and improve our morality, improve our intelligence, and so on, I feel we must do so.

Other objectors to the 'life=defunct' debate

Dr. DMR Sekhar, another debater in the Hmolpedia, took another stance on this debate.

DMR Sekhar: ... *"living systems" are self driven using the internal energy they produce by consuming/ absorbing food/ Sun light from the surroundings. If this is not true then one has to identify the external forces that act upon living systems to move them. You will find no external forces that push you when you are walking. This simple observation should make you rethink. Hope you will realize sooner or later. Kindly take my advice in good faith.*

David: *I think 'non-living systems' as well can have internal energy. For instance the sun's nuclear reaction can be thought of as internal energy. Or friction can cause an increase in internal energy in objects. Internal energy is not exclusive to 'living systems'.*

The sun clearly does not arrive at its energy by extracting it from external forces.

Sekhar: *True physical systems also have internal energy but no "Self Drive" that comes from "self programmability", "consciousness" and "freewill".*

He posts some links to his blog on *dmrsekhar.wordpress.com* .

David: *can you explain what you mean with 'the genome is self-programmable'. I understand from my own reflections that the genome is something might[y] important. you also make some link with quantum physics, can you explain?*
I think, for the most part, I would disagree with you. I agree that intelligent beings have something called consciousness, if consciousness is defined on the psychological level in terms as 'can reflect on its actions', or 'having a wide range of possible behaviors'. consciousness of course may also be related to the brain. but not all living things have a brain. some suggest microtubuli have something to do with quantum computation for consciousness. However, why should you declare the living things as 'self driven' and the non-living things as 'not self driven', if they both have internal energy which can be used to make mechanical energy? a molecule can walk ,as shown on this website, because of chemical laws. a cell will be moving as well, in response to its environment. at higher levels of intelligence, of course, the environment begins to form internal stimuli - the possibility to anticipate. nothing is truly self-driven, consciousness or thoughts come from past experiences. the term consciousness thus is not an absolute.

In short, my theory of 'self-driven-ness' is that it does not exist: external stimuli (for instance light, sound, air) exert force on the individual, to be rightfully called 'external forces'. These external forces cause internal energy in the individual. For instance, when perceiving light or sound, our memory has an internal representation of this visual or auditory stimulus. This 'internal energy' then leads to the power of anticipation, via the process of 'conditioning'. Alike stimuli will be 'recognized'. Recognition is as well a chemical process, perhaps related to dopamine. Hence comes the 'action', or better, transforming internal energy into kinetic energy – all based on the sensory input of past events (which is now present as 'internal energy', since the events are not occurring), and of course the present sensory input.

Then, Libb acknowledges that my view is correct, and points to some interesting pages on the Hmolpedia that show this. Libb sees DMR Sekhar's response as another way of the age-old trying to salvage the free will. Interestingly, the 'fully self-driven-ness' theories are in violation of the combined theory of thermodynamics (a combination of the first and the second law of thermodynamics). An oversight into this - to be found on Hmolpedia as 'perpetual motion' - question is the following:

Energy motions	Entropy motions	Free energy motions
first law +	second law =	combined law
$dQ = dU + dW$	$\int \frac{dQ}{T} \leq 0$	$\Delta G < 0$
Energy is conserved +	Entropy must increase =	Free energy must decrease

The latter column, which equates to the fact that all motions on the surface of the earth must abide to the governing nature of the combined law of thermodynamics, which in regard to human motions, as well as to all animate motions on the surface of the earth, wherein systems are defined as coupled freely-running isothermal-isobaric surface-attached reaction systems, the Lewis inequality for natural processes:

$$\Delta G < 0$$

is the governing equation to motion. Hence, any theory of human motion, or animate motion in general, or device, machine, or molecular structure that claims to be able to produce motion, in opposition to the combined law or system free energy decrease, is what is called perpetual motion of the living kind and is thus impossible.

The conversation proceeds. DMR Sekhar gives some links to his books ("genopsych" and "Consciousness, Entropy and Evolution") and continues:

DMR: *My difference with Libb is that I say that I take my decisions through internal stimulus of (my) will that is my free will. Libb probably thinks that he is not responsible for his acts as some external forces guide his actions. Best.*

Seems to me like another example of someone who says I'm free. Goethe once famously said "None are more hopelessly enslaved than those who believe they are free".

David: *my reasoning is that 'we can choose what we do, but we cannot choose what we want'. in other words, everything one does is determined by previous causal chains. external forces turn into internal energy - nothing magic. the difference lies in how long the causal chain is. For instance, external stimuli create an internal representation of this stimuli in the brain, which can make us anticipate. when we anticipate, and hence we 'choose' what to do, our actions are fully determined by this stimuli.*
By the way, a nice question perhaps is the following: "is the sun responsible for his explosions?" You would have to conclude that the answer is yes. the difference just lies in assuming some person.

The term genopsych centers around DNA (obviously) and consciousness. DMR supposes that there is some quantum computing going on, which supposedly gives special consciousness. This raises some questions:

David: *another interesting question, relating to the following text: "We know that the plane can not fly up on its own spontaneously. It needs a conscious effort by a pilot to take the plane into the sky. Even if it is a remotely controlled as unmanned plane one has to control it remotely and a consciously made programme is always there that automatically controls the flight."*
do you know self-driving cars (they don't need any remote control)? are they conscious then? no genes are needed for this supposed consciousness. thus it seems unlikely that quantum-computing of the dna (or what?) would yield some special results in terms of consciousness.

Of course, we can define consciousness as intelligent, which would mean 'anticipatory power'. This anticipatory power has only an *apparent* relation to free will. Anticipatory power namely derives from memory, which derives from sensory input, and is in theory fully predictable, given complete knowledge of neuronal structure and all the input. Another aspect of intelligence is 'knowledge'. This refers to the byte capacity of the memory. So we have two aspects of intelligence or consciousness if you will: anticipatory power and memory. The former is derived from the latter. The memory of course derives from sensory input. However, how are memories formed? This is one important question! Does this take consciousness? One can store information on small USB-sticks so it seems to require no biological properties (such as genes).

One may define consciousness as computing, but this then leaves us with the question: is this computing voluntarily? not really, the USB stick, or the computer or the self-driving car, accidentally came into being. the human as well, accidentally came into being :-) , and accidentally comes to his thoughts. all just to point out consciousness, if defined as knowledge or anticipatory power, has nothing to do with free will - if we define free will as we have used all along: the opposite of determinism.

A difference might lie in self-programming on the basis of this knowledge? There appear to be indeed some fundamental biological rules: directed toward survival, toward genetic success. However, what causes this directedness? Is it the genes? Since this is indeed a thing exclusive to biology, we can suppose so. Yet, reviewing all of this, and comparing this to a computer, or even a tree, or a piece of metal, we can evaluate a directedness as well. Computer is directed towards the tasks he was programmed to do. Similarly, humans are programmed by their genes, and their sensory input which results in memories programs their thoughts within this genetically determined program. The piece of metal was programmed to be attracted to magnets, one could say. The piece of metal was programmed to 'stay alive' and in intact shape, we could say. Everything is programmed to do something. Whether this something is 'purposeful' is just a matter of cockiness. For instance, metal does not need to 'do' anything to 'survive': it can reproduce by wear and tear into smaller pieces of metal which can be integrated into other pieces of stuff; it can sit still for centuries and still be there. Also, striving for survival, or better, coherence is something that is more determined by the laws of free energy, than by anything else. If these laws dictated that the human molecule would not cohere, than there would be no human. If these laws dictated that an iron plate would not cohere, than producing iron plates will be impossible.

At some point, such a bottom-up debate must turn to a top-down approach. The brain is something special. It can 'think about thoughts'. Thought is the internal representation of stimuli, sometimes connected with other such representations, to form 'abstractions' – the internal representation of things that never existed, but may exist, or are a connection of things that do exist. Clearly, these things derive from causes, indicative of deterministic laws. Nevertheless, they seem to create things that come out of nowhere, and thus at the apparent level, they *seem* completely new, while they are just an integration of many old things.

A further digression on 'knowledge'... How do atoms 'know' how to bond to which atom? Such a question is no doubt very difficult for someone fan of genopsych theories. It seems to be determinate. Going up a scale higher, does the piece of iron know what to do to 'survive'? It seems so.

To give a somewhat practical view on the matter: the 'will to survive' does not really exist; but we can create such a label – after all, with 'abstraction', as mentioned, all sorts of things that don't really exist can be made. This 'will to survive' then is nothing but a manifestation of the laws of free energy – which accidentally gives cohesion of particles in some systems.

However, when going into larger scales, this cohesion can form systems that have a large array of behaviors – of course by means of the brain. This large array of behaviors is close to free will. It is not entirely free as discussed, but nevertheless seems so, to some. Perhaps the mind on its own can and should be considered. Of course, very deterministic laws hold for the mind: input -> storage ->representation. Thus external forces are connected to the workings of the brain. Now, we still have to consider the internal forces in the brain. As noted, abstractions are made by integrating multiple representations. These abstractions can further be integrated into large networks of abstractions. Therein lies the unicity of the brain.

Sekhar: *Kindly note that all apparent "self driving" physical systems are designed by conscious living beings where as the "self drive" of a living system does not require external intervention. See for example " genome is self programmed" which means that the programmes in the genome are not incorporated by any external agency.*

On one point, I have to agree: that the genome is very important, and that it determines what our body will look like. Personally, I'm not a big fan of such terms as 'genopsych', 'self-programmednes'. Also, if something programs itself, then it already existed before it constructed the program to run on. That's quite similar to the statement 'the universe can and will create itself', a statement by Stephen Hawking I will tackle in my other book on cosmology and evolution. Such a statement implies that the universe already existed, which means it cannot 'create' itself. In retrospect, self-programming can exist in the sense that experience recalibrates sensitivities of the brain (to light for instance). How exactly the genome may program itself may be the subject of a later book, where I may treat the subject of Genopsych by Sekhar. More likely, our body is programmed by our genome – and our genome is programmed by a long history of evolution.

Digital philosophy

To what degree abstractions are allowed in bacteria is unknown, and perhaps abstractions itself are hard to quantify in terms of bytes. Bacteria have senses though: for instance, spots of opsin chemical on the cell surface, combined with a transport mechanism (not nerves). Going further down the scale, to what degree abstractions occur in the bonding of two atoms is unknown. One would think that some degree of computation exists even at the smallest levels.

Personally, I think the theory of Wolfram and other digital philosophers, allegedly inspired by Leibniz monadology, to be truthful. On Wikipedia we find:

Digital philosophy grew out of an earlier digital physics (both terms are due to Fredkin), which proposes to ground much of physical theory in cellular automata. Specifically, digital physics works through the consequences of assuming that the universe is a gigantic Turing-complete cellular automaton.

Digital philosophy is a modern re-interpretation of Gottfried Leibniz's monist metaphysics, one that replaces Leibniz's monads with aspects of the theory of cellular automata. Digital philosophy purports to solve certain hard problems in the philosophy of mind and the philosophy of physics, since, following Leibniz, the mind can be given a computational treatment. The digital approach also dispenses with the non-deterministic essentialism of the Copenhagen interpretation of quantum theory. In a digital universe, existence and thought would consist of only computation. (However, not all computation would be thought.) Thus computation is the single substance of a monist metaphysics, while subjectivity arises from computational universality. There are many variants of digital philosophy, but most of them are digital theories that view all of physical reality and mental activity as digitized information processing.

A few positive things can be said. 1. gets rid of Copenhagen interpretation and other non-deterministic philosophies. 2. Monist: as opposed to a mind-matter duality, everything is a bit of a computer – it's just a matter of degree. In this way, what some coin intelligence, is a gradual matter. Computation is a universal property of the universe, even in atoms. If not, tell us how is it possible to achieve quantum computers to fulfill our computational wishes?

The universe as a Turing-complete cellular automaton, by the way, refers to the fact that the universe is assumed to be Turing-complete – indicating that a Turing computer, which works according to a certain algorithm, could simulate it. A different, hypothetical type of computer is the Oracle computer, which can arrive at the solution of any decision problem, simply by guessing – not by a certain algorithm – even if this decision problem is formally undecidable.

Some problems are 'undecidable', which means that there is not one single algorithm which leads to successful results. An example of an undecidable problem is Hilbert's tenth problem. Matiyasevich showed this problem to be unsolvable by mapping a Diophantine equation to a recursively enumerable set and invoking Gödel's Incompleteness Theorem. It seems that even mathematical problems sometimes are 'undecidable'. So, given deterministic rules, we cannot always find one single algorithm. This, again, corroborates the view that our minds follow very deterministic rules, but that these rules are not reducible to one single algorithm – and hence it *seems* we have free will.

One may view the world as a giant computer in which a massive amount of information is stored. In every atom lies such information . Hence, in macrostructures as humans, much information is combined, to give rise to relatively high intelligence. There is no question of free will – everything follows deterministic rules. High intelligence then, means, a large network of computational power. That computational power derives from deterministic laws.

There seems to be a gap between the microscopic determinate laws and the macroscopic 'computationally irreducible' phenomena. Because there is not one single algorithm for these complex phenomenon, the only way to discover how those 'computationally irreducible' things behave, is to *perform* the computation – according to Wolfram. Indeed, this is why the cellular automata approach to science holds much promise.

While the Pythagorean philosophy holds that all is number, we can describe reality with something broader: computation. Indeed, computation is a broader term which encompasses all algorithms, including mathematical.

Intelligence being the extraction, storage and manipulation of information, comes in all things. Information can easily be defined as containing bits. And bits can easily be defined as binary information – the presence of a characteristic or not. On the basis of the presence of a characteristic, even chemical elements distinguish between each other – because of the yes-no property of having or having not a certain amount of electrons in a certain shell, elective affinities exist between chemical elements. In this way, our brain is nothing but a clump of molecules which are able to encode trillions of characteristics in a yes-no matter. Our brain, certainly, is very analogous to a processor, decoding things in terms of yes-or-no. What about molecules? Certainly molecules contain information that can be encoded into bytes, if not, our computational system would not be able to smell them. One can of course, write this into an algorithm, in the lines of *if(molecule1): cout<< 'juk!'* ; *if(molecule2): cout<<'mmh!'*. If one is to maintain the reasonable stance of monism, one must agree that 'everything computes'. If molecules do not compute, then at what point does computation occur? The relative configuration of atoms of course matters in the extent of computation. Still, at some point, computation should arise. My theory is that everything computes – this is one way to explain thoughts from a deterministic standpoint. What is computation? Well, sometimes one defines it as encoding information, usually in terms of yes-no presence of a characteristics. In a sense, everything encodes information, from the sensors of bacteria, to the brain of man. And, perhaps, as well way down the scale on the quantum level, since we can make quantum computers.

Perhaps another opinion can be construed, however. If processors are to be interpreted as the brain, then only given some special configuration, can there be construed yes-no information processes. In this way, the brain and other sensory mechanisms (sometimes the microtubule of the cells are proposed) are a unique construction that allows such computing. In that case, computation is unique to biology and man-made processors and computers.

However, computation then is merely equivalent to some physical principle of registering information (for instance light) and conditionally giving a particular output. This is, I believe, universal. Everything, given a particular input (for instance a beam of light, a neighbouring atom, an electromagnetic field), gives a particular output. So, while everything can be modeled as a computer, information can be equated with chemical processes. Every element follows a certain program. Information storage may be seen as an abstraction, that indicates the amount of things happening (the input) that are relevant for the behavior of the system in question (the output).

Back to the debate

Libb: *Sekhar, brushing away the near inane ideas of yours (e.g. that genopsych is god; that genes have consciousness that moves against the second law; that humans have self-drive, as in perpetual motion; that you think Shannon information theory has something to do with entropy; that you think that humans are alive, etc., etc.), what I don't understand is how you can turn your back on chemical engineering? The following outline is how human relationships are determined, according to chemical engineering:*

[link to page: In thought experiments, the **Thims thought experiment**, a modern-day free energy version of the affinity chemistry based Goethe love thought experiment (1808), refers to the chemical thermodynamics based mate selection thought experiment conceived in circa 1995 by American electrochemical engineer Libb Thims, according to which a person selects mates based on Gibbs free energy change, in the same manner as a physical chemist or chemical engineer would determine the feasibility or spontaneity of any given chemical reaction using free energy tables.]

not by "free will" but by "free energy". In the 1999 words of Canadian-American physical chemist Julie Forman-Kay, "Whether two [people] will bind is [completely] determined by the free energy change of the interaction, composed of both enthalpic and entropic terms." In other words, the chemicals or people have "no" choice in the matter, the reaction is predetermined by the free energy change of the system.
I don't know how you can call yourself a chemical engineer, siding with religion (mythology) over thermodynamics (reality).

Sekhar: *Dear Libb,*
[1] I am aware of your views on genopsych as you had already expressed them earlier. Initially I posted the ideas online for the purpose of Open Review and I am thankful to you for participating in the discussions. Now I finalised the theory taking into consideration all the discussions/ comments and published two monographs on the subject.

[2] You wrote, "In other words, the chemicals or people have "no" choice in the matter, the reaction is predetermined by the free energy change of the system".
People are not just chemicals rather people are chemicals plus. Outcome of simpler reactions may be predetermined but the same is not true when it comes to complex systems. The outcomes of complex systems to the external and internal stimulus are probabilistic! The first principle of Statistical Process Control is "variations are natural". Libb you are ignoring statistical thermodynamics.

[3] What all you are writing about God and other aspects is not a part of my main theory as you know it. It all depends on how you model "God" ! Further discussion on genopsych will be limited to the material published in the monographs. Philosophical interpretation of genopsych is altogether a different cup of tea.
Thanks,

Sekhar

Libb: Sekhar, you are what's classified as an "ontic opening theorist", i.e. someone who tends to go on and on evoking sideline principles to support their slanted and biased view. You are biased in that you believe humans have free will and will continue to invoke theory after theory, e.g. "complexity theory", "process control", "cybernetics", "probability", "statistics", "information theory", etc., endlessly, in support of your position.
Myself, conversely, I accept experimental findings and laws determined therefrom as explanations of my existence, whatever they may imply. In regards to people, experimental measured findings indicate that humans are molecules moving on a surface (Ecological Stoichiometry, 2002) and in regards to the freedom these types of surface attached molecules are allowed, as determined by the laws of physical science, the solution was worked out over 200-years ago, the truncated version of which is written on the sign held by the women in the following photo [photo showing the famous Goethe quote: "None are more hopelessly enslaved, than those who believe they are free"].

This maxim comes to us from Goethe's 1809 Elective Affinities (part two, chapter five)

David: *Can you explain why you don't think information theory is related to entropy? wikipedia, for instance, under information theory gives quite a lot coverage of entropy.*

Libb: *the first JHT article for 2012 is going to be on that subject:*

http://www.humanthermodynamics.com/Journal.html#anchor_240

For the time being, the issue is summarized well by German physicist Ingo Müller as he states in his 2007 book A History of Thermodynamics:

"No doubt Shannon and von Neumann thought that this was a funny joke, but it is not, it merely exposes Shannon and von Neumann as intellectual snobs. Indeed, it may sound philistine, but a scientist must be clear, as clear as he can be, and avoid wanton obfuscation at all cost. And if von Neumann had a problem with entropy, he had no right to compound that problem for others, students and teachers alike, by suggesting that entropy had anything to do with information."

Also it's not a matter of "think", the information theory is about the mathematics of 1s and 0s sent down a transmission line (e.g. telegraph wire), entropy is about the mathematical physics of heat sent into or out of a body (e.g. water in a steam engine). The connection between the two is a result of an inside joke between Neumann and Shannon. The reason for the proliferation of the belief, e.g. such as found at Wikipedia, and everywhere else is ignorance, i.e. science has become very culturally divided, most being nearly incompetent outside of one's own specialty.

See the 2012 chapter pdf handout (section 14.11 Information theory) of my April engineering lecture for further information:

http://www.humanthermodynamics.com/HT_lecture__2012_.pdf

http://www.eoht.info/page/Libb+Thims+%282012+lecture%29

Perhaps in a longer conversation I could have found more common ground with Sekhar. Be careful to draw conclusions with regard to Sekhar's philosophy.

Also, perhaps after more research I could evaluate whether the standard theories -which say that it can be shown in many ways that information entropy (by Shannon) and thermodynamical entropy (by Boltzmann) are equivalent- are true. Not having the time, I will assume they are indeed true – in contrast to what Libb assumes.

5. Other interesting people at the Hmolpedia

Edward Ockham

At the Hmolpedia, under Edward Ockham, we find some interesting comments which he made in a 2012 blog. Added between brackets are comments of Libb.

"Someone [Libb Thims], probably with an education entirely confined to the hard sciences [chemical engineering and electrical engineering], has the insight that people are made entirely of atoms. Molecules are made of entirely atoms, ergo humans are molecules. Materialism of the crudest sort. What's wrong with it? Well, I am not sure it is even scientifically correct. Molecules are arrangements of atoms in certain bonding relationships that hold only at the atomic level. So even DNA is not a molecule, but rather a pair of molecules held tightly together. The relationship that ties the heart to liver, and the liver to the brain is not an atomic one. Or is a molecule a set of atoms in any relationship whatsoever? Then a city is a molecule [social molecule], the Earth [earth molecule] and the Sun [sun molecule] are molecules, the Earth and the Sun together are a single molecule, the whole universe is a single molecule [universe molecule]. That is no help whatsoever."

"Even if it is scientifically correct (I'm no expert), how does the insight help? We want to explain the nature of money, for example. Now money is an arrangement of atoms – either atoms of pound notes, or coins, or bond 'paper', or their electronic correlates in the general ledger of a payments system. But how does that help explain money? It is the job of the sciences of economics and finance to do that. How does the science of atoms and thermodynamics help us here? That's not to say that, once we have perfected those sciences, we could give a more complete, but vastly more complicated explanation in terms of atomic theory. My point is that the insight – that things are composed of atoms – does not help us explain economics, <u>aesthetics</u>, history, etc."

David: Ockham here makes some interesting points, somewhat relating to some of my objections.

first, "Well, I am not sure it is even scientifically correct. Molecules are arrangements of atoms in certain bonding relationships that hold only at the atomic level. So even DNA is not a molecule, but rather a pair of molecules held tightly together. The relationship that ties the heart to liver, and the liver to the brain is not an atomic one. Or is a molecule a set of atoms in any relationship whatsoever?". I'm not sure either, but in any event, it often makes sense to look at particular parts of the 'human molecule' such as the heart.

second, "How does the science of atoms and thermodynamics help us here? That's not to say that, once we have perfected those sciences, we could give a more complete, but vastly more complicated explanation in terms of atomic theory. My point is that the insight – that things are composed of atoms – does not help us explain economics, aesthetics, history, etc." In many cases, overcomplexity is not needed, and is not even of more predictive value. For instance, we don't need to (in most cases) define the hypothalamus or the nerves of the penis in terms of chemistry to predict when erections occur. Or we don't need chemistry when we know that the person's heart stopped beating. Even if we do explain those things in terms of chemistry, we still don't adhere to the human molecular vision, since we are slicing it up in (functional) parts. For such reasons, I see human molecular vision as complementary, rather than the only game in town.

Indeed, science is about predictions, and we must hold on to the approach that has the highest predictive value. For those arguing that 'truth is more important', recognize that truth usually is around when predictions succeed. If predictions fail, then it is clear that there is something wrong. Oh yeah, do I hope that people are not taking interest in the subject of 'money thermodynamics' – precious effort wasted.

Third, not really an objection, but rather an interesting note, is the piece on "Therefore what I did was not wrong". Einstein once supposedly said (dixit Michio Kaku) that everything was determined, and thus criminals couldn't help their actions, but 'we should still put them in jail' :).

An interesting comment by Libb on the blog article is the following (to be found again on the Hmolpedia page on Edward Ockham):

Regarding his second comment: "we can characterize thought and understanding entirely without reference to any external object", this is incorrect logic. To exemplify, adjacent is the animate 3-element retinal molecule, which in descent down the evolution timeline is but a smaller, albeit less complex, version of the animate 26-element human molecule.

Indeed, it seems aspects of cognition are dictated by external factors: sunlight for instance comes through the retina and determines our wakefulness. Edward Ockham nevertheless has made an interesting point, that we *can* describe the mind without external factors. In my opinion, we should emphasize the *can* (I'm not sure whether he intended it this way): it is one of the possibilities. Given that we can keep external factors (relatively) constant, it is one approach to ignore external factors, and only study the mind itself. This is what cognitive science, and AI science does, in many instances.

Libb: Re: "I am not even sure if it is scientifically correct, etc.", the issue here is that we are venturing into a new age of science, wherein boundaried systems showing a free energy differential decrease is the new overriding rule or "supreme law", as Eddington put it, governing natural processes:

http://www.eoht.info/page/Supreme+law

In other words, the overriding rule of changes in human organization and existence is changes in free energy:

$G = H - TS$

on going between two different states of existence. Say, for example, the United States, as a system, has a free energy value of 35 gigajoules per hmol in the year 2020 and that two different intrepid experimentalists, hmolscientist #1 and hmolscientist #2, determine or calculate two different year 2090 free energy values: #1 calculates a value of 65 gigajoules per hmol and #2 calculates a value of 15 gigajoules per hmol. Based on on the universal rule of thermodynamics (or supreme law as Eddington puts it), version #1 is not thermodynamically possible, hence we would rule out their version of the calculation. In all of this we need to think of the system of the united states as a "chemical system" reacting over the course of those 70 years. This is all new science. Hence, we need to fall back on first principles, namely the original definition of molecule by Pierre Gassendi as a "structure of two or more molecules." In this perspective, DNA can be defined as either one molecule *or as two molecules, depending on point of view:*

http://www.thenakedscientists.com/forum/index.php?topic=37150.0

wherein the thermodynamic point of view is the overriding one. Likewise for a married couple, which can either be defined as one molecule, a dihumanide molecule:

http://www.eoht.info/page/dihumanide+molecule

or two human molecules attached via a chemical bond. This is what is called "big science".

David: *thanks. there are certainly many hints to the importance of thermodynamics. the dihumanide molecule only refers to a couple that is close together, right? or how can there be an atomic bond at large distance - spooky action at a distance :D .*

Certainly, many have stressed the primary nature of the second law of thermodynamics.

I'm still not sure whether we can accurately say that big structures like humans, the earth, a spaceship, etc. can be classified as a molecule. For instance, if we have a closet made out of parts, with most parts only loosely attached, can we then say that the closet is a molecule? Those bonds are not an atomic bond, but rather a spatial vicinity – is my guess If we have a bottle of salt, can we say that there are bonds with the glass molecule and the multitude of salt molecules? Similarly, if we view the human as one molecule, this perhaps somewhat ignores the fact that there are many parts, that have no atomic bonds with each other – for instance, urin in the blatter or blood in the veins is more floating around than bonding to form a large supermolecule. Of course, we may also define molecule as 'some bounded system' – in which case there would be talking of a 'human molecule'. So, again, it's a matter of definition.

It certainly is an interesting new approach to the problem, which can perhaps discover things that are otherwise impossible to discover – and therefore it's interesting to stimulate this kind of research. Also, this research starts from the primary principle of existence – thermodynamics.

Some objectors to the human molecule

Besides the comments of Edward Ockham, I found other interesting objectors. These objections are to be seen at Hmolpedia under 'libb thims (attack)'.

A few people made comments that lie within my reasoning.

Philip Moriarty : *Where did Gibbs say that 'a society is one such material system'? He didn't –that is your particular (incorrect) reading of the application of thermodynamics.*

When I talk about the 'entropy' of the students, it's **really** important to note that this is just an analogy. Entropy and the second law of thermodynamics are very abused concepts.

An arrangement of the students *does not* have an associated thermodynamic entropy.

You have taken the abuse of the term entropy to an entirely new level, however, by suggesting that it—and, unbelievably, quantum mechanics—can be applied to 'interactions' in romantic human relationships." "

From many of my investigations, I conclude that entropy is almost as badly defined as is life. There seems to be little consensus…

Also the suggestion of the human molecular orbital is probably what Moriarty refers to with quantum mechanics. He is probably right that its relevance to romantic relations is false, and that it is unlikely that quantum mechanics can be extrapolated up to the human level. Still, Harris, for instance, has shown that quantum mechanics also apply to larger systems, up to about 1000 atoms. I would consider it plausible that it is as well applicable to human beings – although it remains unlikely that an orbital would be indicative of 'love'.

Georgi Gladyshev: *"Hydrogen molecules and human molecules are objects of different categories. Your viewpoints about the general nature of 'your molecules' contradict to the philosophy, to the principle of holism.*

Indeed, if human molecules consists of different things, have different properties, then many analogies will fail. Perhaps some will succeed. Let's find out.

Lubos Motl: *"Human beings are NOT molecules, they are composed of molecules, but we aren't giant molecules."*

Indeed. Unless we redefine molecule – see final chapter.

6. Morality

Libb has stated : "I believe in the atomic theory and thermodynamics and this is the one-sided basis of my belief system. Both Newton and Einstein corroborate with this belief system, as does Goethe, but Goethe does so to a greater extent. Humans are indeed "merely physico-chemical compounds" as has been proved in 2002 [He's talking about the 'human molecular formula]."

The fact that there are physical laws to which everything existent is subject, is not new and is widely accepted. However, concluding erroneously that "merely physico-chemical compounds" cannot lead to laws defined on other levels is not what this determinism implies. It is clear that our brains, determined by all of the memories and chemical reactions occurring, still are key to explaining our behavior. So studying the brain is much more interesting than to study 'human molecular marriage bonding', which is basically the same as saying – oops we attract, so now we are a married couple – in reality there is much more to it, such as social rules, moral code, rituals and so on. Mere attraction is not sufficient to explain 'marriage', which is for all a social and cognitive institute.

Libb will also refer to concepts as 'good' and 'bad' as inexistent. I can follow his reasoning in the sense that good and bad are not defined in terms of physical laws. Libb also likes to comment on the often anthropomorphizing terms I and others may use, such as 'life'. But then it strikes the vigilant reader that Libb himself has made a paper, which supposedly proves, thermodynamically, that good wins over evil. In the paper, Libb defines good as natural: $dG<0$; bad as unnatural: $dG>0$ – where dG stands for Gibbs free energy. I can't see any more direct anthropomorphizing of a law that has nothing to do with morality. Morality exists as rules for intelligent beings. It does not exist for the laws of physics. Morality is a set of rules to make interaction of intelligent beings more pleasant.

Morality is certainly a point where Libb often proposes that it should be done from a perspective of determinism, of chemistry and so on. But clearly, morality is not defined at the chemical level – it clearly is a set of relatively arbitrary agreements that are made to be happy and to get along; to avoid pain. Our rules of morality, which overall I agree with, are based on achieving happiness and understanding.

The morality that Libb would propose, is never explicitly proposed. Rather, Libb, each time after he says 'life does not exist', he goes on to claim that this should have something to do with morality. To me, the most obvious moral principle that would follow from 'life does not exist' is that 'it does not matter whether we would kill somebody', since life does not exist. Clearly, such reasoning is highly *immoral* and I hope Libb does not propose this.

So I decided to ask in detail how Libb connected our debates to morality.

Libb: *The morality page is here:*

http://www.eoht.info/page/Morality

The gist of all morality theories have to do with differentiating the sum total of human movement into two types: good and bad. The dominant morality theory, which three fourths of the world adheres to, derive from the negative confession theory:

http://www.eoht.info/page/Negative+confessions

namely that each "confession", e.g. "I have not stolen" (#3), has a certain measurable mass associated with it, the sum of which determines one's fate in the afterlife.

Most of all of this we can jettison, except for the reality that we do "move" and that we know that thermodynamics differentiates two types of movement: natural and unnatural:

http://www.eoht.info/page/Natural

http://www.eoht.info/page/Unnatural

and that we know that these movements are "coupled" together:

http://www.eoht.info/page/Coupling

This is where the upgrade needs to occur. This is what Goethe was pointing towards, 200-years ago, when he said that the "moral symbols" in nature are found in the physical chemistry textbook. This is what Victoria Woodhull meant when in her 1871 summary of Goethe's Elective Affinities, she said "A great revolutionary doctrine pervades the whole." Overthrowing all the world's religions, however, is no simple matter.

I decide to take the first page to discuss, since I find this will be quite sufficient to start with. I will show separate chapters of the page, and will discuss each one at a time.

Physical science morality

The modern-day physical sciences based model of morality is that (a) a person is animated reactive 26-element molecule (human molecule), (b) that "good" actions, or rather "natural" actions, are governed by the Lewis inequality for a natural process (dG < 0), (c) that "bad" (or evil) actions, or rather "unnatural" actions, are governed by the Lewis inequality for an unnatural process (dG > 0), (d) that both natural and unnatural processes are thermodynamically "coupled" together, such that natural processes energetically drive the unnatural processes and that some reactions will progress in a direction contrary to that prescribed by their own affinity, (e) that there is no such thing as "life" and "death" according to the defunct theory of life model. This is the thermodynamic explanation to age-old idiom that "good always triumphs over evil", which means that natural processes will always triumph over unnatural processes or technically that the total set of processes will only go when the system shows an entropy increase or transformation content increase.

The details of this model, however, are largely underground and remain to be ferreted out to the general public. In 2010, American neuroscientist Sam Harris, in his attempt to found a deity-free morality system in pure science, asked, for instance: [8]

"In a world of physics and chemistry, how could things like **moral** obligations or values really exist?" In sum, the modern answer to this query is the Lewis inequality, which distinguishes between what is 'natural' and 'unnatural' in the course of human existence, viewed in the context of thermodynamic coupling and universal spin, spin being what drives the various heat cycles of daily existence.
To make this more readable, we have to include the chapter on coupling.

Affinity coupling

In the years 1920 to 1936, Belgian physical chemist Theophile de Donder worked out the basics of "affinity coupling", such that if in a system two simultaneous reactions occur:

$A_1v_1 < 0$
$A_2v_2 > 0$

where A and v are the chemical affinity and chemical reaction rate, respectively, and the subscripts refer to reaction one and reaction two, respectively, that the overall reaction will occur as long as:

$A_1v_1 + A_2v_2 > 0$

at which point the reactions are said to be "coupled" reactions. In this sense, according to de Donder's protege Belgian chemist Ilya Prigogine, "thermodynamic coupling" allows one of the reactions to progress in a direction contrary to that proscribed by its own affinity. The rules for affinities of reaction are defined as follows: [5]

Measure	Description
$A > 0$	reaction proceeds to the right
$A < 0$	reaction proceeds to the left
$A = 0$	reaction is in a state of equilibrium

The formula de Donder employs to measure affinity A is:

$$A = -\sum_{i=1}^{j} \nu_i \mu_i$$

where μ (mu) is the chemical potential, as defined in the work of Willard Gibbs (1876), of one associated molecular entity or mass ν (nu).

Free energy coupling

In thermodynamics, "free energy coupling theory" is a relatively new theory, first outlined in terms of Lewis thermodynamics in German-born American biochemist Fritz Lipmann's famous 1941 "Metabolic Generation and Utilization of Phosphate Bond Energy", which itself was based on previous scatters works on the puzzle as to how to explain the energetics of frog leg movement. Lippman's free energy coupling model is summed up by the following equation:

$$\sum_{i=1}^{j} \Delta G_{N_i} + \sum_{i=1}^{k} \Delta G_{\tilde{N}_i} < 0$$

is called the Lipmann coupling inequality and states that as long as the sum of the Gibbs free energy changes for all "natural", symbol N, processes or reactions, in a given coupled system, plus the sum of the Gibbs free energy changes for all "unnatural", symbol $\tilde{N}$, processes or reactions, in a given coupled system, is negative or has a measurement value less than zero then the process as a whole will be natural and spontaneously progress or occur.

In modern times, this is known popularly as the "free energy coupling" model of driven energy transformations, the textbook example being the model of ATP as a type of "energy currency".

A later spin-off the coupling model is English biochemist Peter Mitchell's 1961 chemiosmotic theory, or chemiosmotic hypothesis, the theory which explains the coupling processes that made ATP in the first place as well as the thermodynamics of membrane transport in general.

Out of the natural tendency for Gibbs energy to go to a minimum, a quantitative measure as to how near or far a potential reaction is from this minimum is when the calculated energetics of the process indicate that the change in Gibbs free energy dG is negative. This means that such a reaction will be favored and will release energy. The energy released equals the maximum amount of work that can be performed as a result of the chemical reaction. In contrast, if conditions indicated a positive dG, then energy—in the form of work—would have to be added to the reacting system to make the reaction go. [end of excerpt from the Coupling page]

[I pick up on the Morality page, again starting from Goethe] In 1809, German polymath Johann Goethe introduced his purely physical chemistry based 'moral symbols' theory of morality.

Systems of **morality** refers to ways of theoretically describing "right" and "wrong" types of human behavior, actions, and or modes of conduct.

A thermodynamics structured update to Kant's categorical imperative, was German physical chemist Wilhelm Ostwald's 1912 'energetic imperative'. Ostwald seems to have been one of the first to state his morality system in mathematical terms:

$$G = k(A - W)(A + W)$$

where G is Gluck (happiness), A is Arbeit, German for 'work', referring to energy expended in doing useful work, W is Widerstand, German for 'resistance', referring to energy dissipated in overcoming resistance, and supposedly k is a constant. Ostwald's energetic imperative, in the century to follow, led to various similar 'thermodynamic imperatives' outlined by various authors.

Into the 1970s, various writers began expounding on verbalized conceptions of 'entropy ethics'.

In 2010, American neuroscience philosopher Sam Harris introduced his idea of deity-free 'moral landscape'.

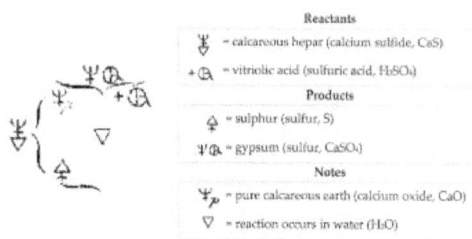

Goethe

The foremost historical attempt to quantify morality in terms of pure science is German polymath Johann Goethe's 1808 theory of people viewed as large evolved versions of reactive chemicals and morality explained in the symbols of chemistry, such as letters used to represent single reactants, A or B, the reaction arrow →, signifying change, and the chemical bond AB, signifying a union, etc., and the energetic measure of the force of the reaction, namely chemical affinity A, defined in modern 1882 formulation as

Example of some of Swedish chemist Torbern Bergman's 1775 "moral symbols" according to German polymath Johann Goethe, which can be used to explain human morality.

$$A = T\Delta S - \Delta H$$

being a function of temperature T, entropy change ΔS, and enthalpy change ΔH.

In sum, Goethe was the first to state that morality is based in the logic of chemistry and physics was

Depiction of a "**moral compass**" a device, analogous to a electromagnetic compass, that can give direction to paths of good vs evil and progress vs regress.

German polymath as famously encapsulated in his 1808 statement, made a year prior to the publication of his famed novella *Elective Affinities*, as commented to his friend Reimer:

"The moral symbols used in the natural sciences were the elective affinities discovered and employed by the great Bergman."

The adjacent diagram shows Swedish chemist Torbern Bergman's chemical "symbols" (proto-types of modern chemical reactions), from his 1775 textbook *A Dissertation on Elective Attractions*, with which Goethe not only used to explain what is moral or amoral in human existence, particularly in regards to marriage and divorce, but, in a seemingly effortless manner, scripts a complex novella love rectangle over this logic. In a modern sence, Goethe's statement translates to an effect that what is moral or amoral in human activity is a perspective determined according to the free energies of reactions between people.

Important to discuss here is the Elective Affinities by Goethe, which is summarized by Libb as follows.

Elective Affinities

See main: Elective Affinities (book), elective affinity

Torbern's *Dissertation* functioned as the reference textbook for the scripting of the first human elective affinity reactions, found as "layers of Gestalt", in German polymath Johann Goethe's 1809 novella *Elective Affinities*, the founding book of the science of human chemistry. [4] To exemplify this, in 1808, a year prior to the publication of his novella, Goethe commented to his friend Riemer: [6]

"The moral symbols used in the natural sciences were the elective affinities discovered and employed by the great Bergman."

In a conversation with Riemer on 24 July 1809, Goethe specifically names Bergman as a source for his for the idea of "elective affinity", using the early German translation of that term, namely "*Wahlverwandtschaft*", which became an inspiration for his novel. [7]

In his novella, Goethe built on this premise by laying out this logic of "moral symbols", in literary form, where each chapter, in underlying theoretical basis, was considered as a different type of affinity reaction or chemical reaction in the modern sense. In chapter three, for instance, in the mind of Goethe, the Captain *C* arrives to stay with the married couple Eduard *A* and Charlotte *B*, after which time Edward and the Captain rekindle their old friendship, thus displacing Charlotte from their activities. This is depicted below in the Bergman reaction diagram style of what was called a single elective affinity:

I think the following is important:

Or in a modern reaction sense, chemical species *A* and *B* are attached in a weakly bonded chemical union, signified by the bonding bracket "{", ordered such that if species *C* were introduced into the system, the greater affinity preference of *A* for *C* would cause *A* to displace *B* and to thus form a new union with *C*, which equates to the following in modern terms:

$$AB + C \rightarrow AC + B$$

Goethe scripted each chapter in his novella based on variations of these types of affinity reactions.

This seems to make sense from an intuitive point of view.

Ostwald
This physical chemistry basic morality model was worked on further, albeit superficially, by German chemist Wilhelm Ostwald and his 1912 *The Energetic Imperative*; a logic which in turn blossomed into number of different so-called verbalized "thermodynamic imperatives" in the decades to follow. [10] Most of these, however, end in nothing but quaint excursions about order and disorder in the social domain. The deeper nature of morality, initiated by Goethe with his statement that "the moral symbols used in the natural sciences were the elective affinities discovered and employed by the great Bergman", naturally enough, leads into the science of the human molecule and the study of how to understand and model changes in enthalpies and entropies involved in human chemical reactions, which in its current form is still a nascent subject with many issues to be resolved. The central issue revolves around a reformulation of Ra theology, the dominate morality system of the world, in terms of pure chemical thermodynamics, which is a daunting task to say the least.

Let's find out what Ra theology has to do with this issue.

Morality

The system of morality in Ra theology is based on the "negative confessions", which were 42 bad, evil, or forbidden actions prohibited to the Egyptian people, theorized to be morally wrong. The accumulations of these evils in the "ba" or soul of the Egyptian, theorized to be located in the heart, would be transported to the great judgment hall (in the heavens), proceeded over by 42 gods (one for each sin), along with other gods such as Osiris (Ra's great-great grandson), and weighted on scale against the feather of truth, and if his or her soul was found to be too full of sin, they would not be granted eternal life (or reincarnation, in the Hindu-based versions). in the which the question of what constitutes "good" actions, which boil down to specific second-by-second movements of the human molecule, viewed in the structure of the long-term dynamical movement of the universe, is a very difficult topic. Confession or sin number eight (above), for example, states focuses on "lying", where it is said that all lies are sinful. In modern philosophical debates, however, the "concentration camp lie", i.e. telling a lie about something in order to save your own life or the lives of your fellow inmates, is often used as an example of when it is supposedly "good" to lie.

An example stab at this issue, pictured above can be found in the 2009 work of English biophysicist Mark Janes and his attempt at formulating good and evil in terms of enthalpy change ΔH and entropy change ΔS.

It seems really that Libb proposes to jettison all behavior rules and instead proposes us to behave 'naturally'/'good', as according to dG<0. However, what exactly is it that we can do to act moral? Exactly, nothing. The laws of physics are determined, so there is little we can do to change them. So why does it matter to have a moral system anyway? If moral is not something we can change in our behavior, then we cannot, strictly speaking, 'better' our behavior. Maybe this is what Libb proposes: we should not care at all about superficial rules, but rather let the laws of nature reign. But of course, this is no moral at all, since the laws of nature always reign. In this case there is no mode of conduct that is 'more moral' than any other, unless the most basic laws of nature tell us to. So the morality proposed then is to simply neglect all moral rules – can Libb be a serial killer if it doesn't really matter since 1. 'life does not exist, so you cannot remove it'; 2. 'if the negativeness of dG tells me that killing many people is ok, then I must do so.'

Libb may say to me that my utilitarian thinking does not equate with physical reality. However, terms as 'good' and 'unnatural' do not equate with physical reality either. Good is a metaphysical construct based on opinions on things we believe are 'fair' to others in the sense that we would not like these things to occur to us. It is based on altruism, while looking within, thinking 'would I feel pleased in this situation?'. This then is based on emotions, which again is a metaphysical construct. If Libb means with good, simply that dG<0, then why do we need the term good ? Why should we even say that 'natural' is 'good'? Clearly dG<0 is the most natural state (if we compare dG<0 and dG>0). However ...

A 1999 summary of "**godless morality**" systems by English bishop Richard Holloway on the weaknesses of divine authority based morality systems. [11]

Dreier
In 1948, American author Thomas Dreier's 1948 discussed morality in the context of pure objective chemical reactions, i.e. human chemical reactions, divorce, and marriage: [5]

"The trouble is that too many people get chemical reactions all mixed up with **morals**. They call immoral what is only a normal chemical reaction."

If this is really the point, then it is indeed a very tautological philosophy. It has no use whatsoever: of course, we are all chemically reacting things. Why should this imply we can have no moral codes? If you really have no moral code, then you should live up to it. Yes, it is simply impossible, since our moral rules determine what range of behaviors you're restricted to. These moral rules of course, stem from a long history of chemical reactions, well duh. That goes for all things. Again, as with the life debate, it seems that we should dismiss any concept that *derives* from chemical reactions, by simply saying 'It is just a chemical reaction.' So, what do we gain with this approach? I decided to gain more clarity.

David: *For clarity, is your proposal that we should not care about rules such as 'don't steal', 'don't murder', 'don't etc' ; and that we instead should look at in what cases dG > 0 and avoid this? How can we avoid this? (If we can't avoid this, then this approach is of course tautological)*

Another read that surprised me was the following:

Moral monkeys
In circa 2006, experiments showed that monkeys have an inherent sense of some sort of morality. Specifically, in experiments conducted by Dutch primatologist Frans de Waal, monkeys were first trained to pull a lever to get food. Then the lever was hooked up so that when the monkeys pulled the lever to get food, it not only produced food, but severely shocked a monkey in a neighboring

cage. It was found that the monkeys would voluntarily choose to starve themselves, going between five to twelve days without food, rather than shock their neighbor. [6] The extrapolation of this finding is that what we define as "moral behavior" must have its origin in the hydrogen atom, being that humans (26-element molecules) evolved from monkeys (24-element molecules) which evolved from the hydrogen atom.

To be added to the list of the wild extrapolations with 20 levels of difference, is this moral behavior extrapolation: if a monkey can show 'moral behavior', then so can a hydrogen atom… Maybe if you change the definition of morality by saying morality is a law of physics… In this case everything, is at most of the points of its existence, moral – simply because $dG < 0$ will be the most frequent state. The question of 'what is moral?' in that case becomes obsolete.

Moral, again, can be best conceived of as metaphysical, since it has no clear physical referent. The moral rules are construed by thoughts. Just as the meaning of what I am writing has no physical presence, moral rules have no physical presence. They do not contradict the laws of physics of course, since they have arisen via physical processes. In some cases it seems they are physical, written down, talked about etc. But it is not the sound, nor the light waves that interest us, it is its meaning. And this meaning can only come by mediating these light or sound waves via neural sensors.

Unnecessary to make the case more complex, we speak of metaphysical concepts when dealing with thoughts, emotions, and hence as well moral rules. We could agree that thoughts is the larger category, containing both emotions and moral rules. This because of the similarity between emotions and thoughts noted somewhat above, and because moral rules are, similarly, to these two, internal representations of experienced stimuli.

Further on morality with Libb

David: For clarity, is your proposal that we should not care about rules such as 'don't steal', 'don't murder', 'don't etc' ; and that we instead should look at in what cases $dG > 0$ and avoid this? How can we avoid this? (If we can't avoid this, then this approach is of course tautological)

Libb: To give you a short answer, yes instead of turning to the Bible (or Koran, etc.) for guidance and understanding in times of difficulty or moral confusion, we should be turning to the physical sciences, chemical thermodynamics in particular; and yes free energy differentials are where the solution lies. You can glean a bit of this wisdom in the mind of Frederick Rossini:

http://www.eoht.info/page/Rossini+debate

What we are doing here, at Hmolpedia, is working on the problem, rather than preaching proscriptions, as you seem to be intuiting. In the famous 1971 words of Nicholas Georgescu-Roegen: "manifold avenues open up almost as soon as one begins to tackle the problem." Notice that we are up to nearly 3,000 Hmolpedia articles alone in researching these different avenues. Take a look at Mark Janes's video on morality and thermodynamics to give you an idea:

http://www.eoht.info/page/Mark+Janes

So, in regards to your query "is your proposal that we should not care about rules such as 'don't steal', 'don't murder', 'don't etc' ; and that we instead should look at in what cases dG > 0 and avoid this?", correctly, what we are doing here is upgrading everything, especially our belief systems and points of view in regards to right and wrong or natural and unnatural as thermodynamics sees things.

Belief systems are what determines one's actions. Take for example Australian moral philosopher professor Peter Singer, professor of "bioethics" at Princeton, the so-called godfather of the animal rights movement, who objects to humans eating animals, but not to humans having sex with them, a logic based on his speciesism views. He's what's classified as an intellectual moron:

http://www.eoht.info/page/Intellectual+moron

Hence, to your comment "How can we avoid this? (If we can't avoid this, then this approach is of course tautological)", here we are working out the correct belief systems, based on thermodynamics, and putting them down on paper, hence, by natural reaction, there will be fewer Peter Singer type belief systems in the future and more belief systems based on the moral symbols of physical chemistry as Goethe so wisely prophesied 200-years ago.

At the moment, coupling theory is the biggest hindrance to clearer understanding.

David: It sure is an interesting thing to think about. Perhaps you can clarify why dG<0 would mean 'good'. Is this because there is no extra energy needed that these are considered more 'natural'? It's a mind-bending thought of course - that we should change our morality system. Personally, I think it's all about reducing suffering of beings that can feel suffering. Maybe this is automatically implied by this particular reasoning? That's a long leap of course. I think we can all agree that murdering causes suffering, or that wounding does the same, that stealing causes suffering and that raping causes suffering. Perhaps these classes of acting are unnatural. If they are found to be natural, I would suppose that 'natural' does not at all mean avoiding suffering. In that case, I would not subscribe to this morality.

Libb: Re: "Perhaps you can clarify why dG<0 would mean 'good' ", the terms 'good' and 'evil' are anthropomorphic labels, that are not terms found in chemistry textbooks, but are ones based purely on point of view (e.g. Muslims thinking Christians are evil / Christians thinking Muslims are evil) . There are many instances were previously conceived notions of right and wrong become flipped, when new circumstances arise, e.g. viewing steeling as good when destitute, but evil when affluent. In any event, here's a similar response I posted last year to similar 'what is evil' question from my YouTube channel:

http://www.youtube.com/user/HumanChemistry101/feed?filter=1

"To understand good (natural) and evil (unnatural), you first have to view all individual human movements as chemical reactions. Some reactions we call "good", such as love at first sight: Man + Woman → Man≡Woman. These are quantified by negative Gibbs free energy changes. Some reactions we call "bad", such as an unwanted arraigned marriage, or "evil", such as reproduction from farther-daughter incest: Father + Daughter → Baby. These are quantified by positive Gibbs free energy changes. Both reactions, good and evil, occur in society because of what's called thermodynamic coupling.

A simple example of what we might call an "evil" process or reaction would a hydrogen molecule H2 naturally or spontaneously splitting apart in the atmosphere to form two hydrogen atoms: H2 → H + H. This reaction, which never occurs on its own, has a free energy change measure of +157 kilojoules per mol. This is called an unnatural reaction, from the earth-surface system point of view. This reaction can be made to occur, however, if it is coupled energetically to another stronger reacting system (such as a strong battery) or if the system is heated (such as the early universe is hypothesized to have been). In the case of thermodynamics coupling, we say that unnatural reactions can be made to go if they are coupled to natural reactions, as long as the later are more powerful than the former. Hence the motto: 'good always triumphs over evil'."

Generally, to find an objective position on questions such as this, what is need is deanthropomorphism:

http://www.eoht.info/page/Deanthropomorphize

David: The thoughts on coupling are very interesting. However, there are several pieces which I find problematic: the incest thing for example. Have we validated experimentally that father+daughter-> baby reactions are quantified by positive Gibbs free energy changes? Or that love at first sight by negative ones? How would one go about to measure such a thing? I have the feeling that you have just used the method we all use for determining morality – see what would be harmful, or 'unnatural' in the popular opinion- and consequently assign the labels (un)natural and its dG equation.

Should we use a free energy measurement device at all times to guide our actions? If not, how can you tell whether what you are doing is 'moral'? It seems Libb has, post hoc, assigned the dG inequalities with the things we in general find 'unnatural' or 'evil'. Similar seems to be the case with the labeling of intellectual moron, because the person in question decided to consider eating meat as immoral and having sex with animals as not morally pervert: how have we related this to a dG inequality? I suppose we haven't.

Chapter 7: Final thoughts

On the human molecule and the orbitals – how do we define the system

Some things that are in doubt to me regarding to the human molecule -and consequently, my definition of life- is the following.
Say we define life as 1. Having cells 2. Having genes coding for its proteins. How do we define the 'its'. That is, where is the boundary of this system drawn?
One analysis may lead into thinking that the human really is a 'molecule': if we acknowledge that the human is a bound-state, that is, a system comprised of many pieces that are drawn together because of photon-electron bindings. This seems to be the only way to correctly explain why things cohere in such a neatly clumped-together, particle-like structure – the human.

How else could we explain the cohesion of the system? Well, the cohesion of the system itself most likely has to do with the afore-mentioned. I can find no other reasonable explanation for this. So, reasoning that molecules are defined as these types of bonds it may be very well possible that the human can indeed be considered as *one* molecule.

However, again, it is a matter of definitions. Atoms and complexes connected by non-covalent bonds such as hydrogen bonds or ionic bonds are generally not considered single molecules.

But, say we redefine 'molecule' as a structure of which particles are held together via a bound state. The particles that constitute the molecule are subject to a potential such that the particle has a tendency to remain localised in one or more regions of space, this via photon-electron bindings. The potential may be either an external potential, or may be the result of the presence of another particle. Using this definition, the human is a molecule.

I asked Libb's help for the question whether photon-electron bindings are the most general type of atomic bonds, comprising both covalent and non-covalent bonds:
Libb: "why photon-electron? is this the general type of binding interactions, comprising all of the bonds (covalent and non-covalent)?", yes. In the concise 1985 (QED) words of Richard Feynman:

"I would like to again impress you with the vast range of phenomena that the theory of quantum electrodynamics describes: It's easier to say it backwards: the theory describes all phenomena of the physical world except the gravitational effect, the thing that holds you in your seats (actually, that's a combination of gravity and politeness, I think), and radioactive phenomena, which involve nuclei shifting in their energy levels. So if we leave out gravity and radioactivity (more properly, nuclear physics), what have we got left? Gasoline burning in automobiles, foam and bubbles, the hardness of salt or copper, the stiffness of steal. In fact, biologists are trying to interpret as much as they can about life in terms of chemistry, and as I already explained, the theory behind chemistry is quantum electrodynamics."

See Human Chemistry (2007), chapters 6-9, for more on this.

This illustrates the huge importance for seeing the human as a molecule, in the extended definition – and of course the huge importance of quantum electrodynamics.

From this light, the human molecular orbital theory seems plausible, but whether its effect has anything to do with love is unclear to me, although it intuitively has much appeal to say that two human molecules, or a family of human molecules, can bond in the same type of way. The maximum the human molecular orbitals can explain is the following: attraction, an instinct of togetherness. The instinct of sympathy, as a primary factor, resulting in empathic behavior or a potential leaning toward this or that person. Suppose this maximum the human molecular orbitals theory can explain: in that case, our behavior is determined by inclinations toward certain people. Sympathies toward some people will develop, and this will influence our cognition directly (through interaction) and indirectly (the memories and dreams determined by these interactions). The energy that is needed to unbond the humans in question then, comes in the form of mental effort – it seems indeed sometimes difficult to get away from people you sympathize with.

For the definition of a biological organism, this seems to be a handy out-come: we now understand how to make the distinction between being and surrounding. Thus a sound definition of life would be 'a structure of which particles are held together by photon-electron bindings such that they remain localized – in other words, a molecule – that has genes coding for its proteins and its structure, and which consists primarily of cells'.

Another way to solve the being-surrounding distinction would be to just state the rule: 'it is where the cells, and its programmed structures, end'. What I mean with this is that the things that make the distinction - between biological being and surrounding – usually is the place where the cells, or its derivatives (hair, teeth, ivory, bones) end.

It is clear that the genome of the organism programs what becomes – in terms of cells and organs, but also in terms of the products of those cells (think of enamel crystals and bone minerals). Biology cannot be considered as anything different than the program of the genes: our genes determine what we become; any error will result in a serious deviation.

However, this says little about the actual nature of the bond that keeps together the human particle. It is perhaps wise to console the human molecular vision with the notion that DNA is an important notion that determines what this human molecule will look like. As said in earlier chapters, we consider the human molecule a term as variable as humans really are – that is, we cannot synthesize the general human molecular formula and expect there to be a genuine human. More likely, it is the case that synthesizing the calculated human molecule will result into a heap of masses. Without DNA, we would indeed have a tough time creating a human being from scratch – perhaps because we don't know the nature of every single atomic connection. And of course we don't have to: we have to know the nature of the genome only to create a full-blown human.

In short, the human molecular vision seems to be correct, but only if we expand the definition of molecule to include non-covalent bonds. Additionally, the boundary of a biological being can be set where the cells or their derivatives end. But we need to use the notion of bound state to explain that our biological whole coheres in such a neat way. Recognizing that the human can, under slight but handy revision of the notion 'molecule', be considered as one molecule, we still hold that DNA is key to creating a human: it is DNA that determines our exact molecular formula; we cannot simply synthesize a human molecular formula and expect to see something meaningful out of it.

A final conclusion

As with all of science, different methods will do different things. We notice that the medical sciences have given us longer lives, that computer science has given us many ways to communicate, and that physical sciences have brought us its unique technological applications and of course intricate insight into nature itself. We cannot say one is more correct than the other, since *truth works*.

I do not think there is one correct way to define a certain thing. From this light, the human is and is not a molecule – definitions change over time and may change (for the better perhaps?) to include non-covalent bonds. With the definition of life, we have seen that there are a lot of ways life is and can be defined. The most important thing, in terms of definitions, is that we, before we speak, make clear which definition we are using. But most of all, when I say many definitions are possible, I refer to the different definitions given by us through different sciences. We know the definitions in terms of functions – which I deemed effective definitions- which are frequently encountered in the medical sciences. But we also, in this book, have come to know a different methodology. Human thermodynamics. The reader may take the challenges into account, but at the same time recognize that there have always been many challenges to *any* framework that attempts to explain human behavior.

There are many conjectures this science makes. Some of them seem plausible, or even, very likely. The human molecular orbitals theory, in retrospect, seems to be truthful in the sense that our location can be traced into a 90% probability cloud. But, firstly, we could do this theoretically speaking with anything, even if the theory wasn't correct. To make the theory credible, at least one assumption needs to be conquered. The wave-particle duality must hold for very large molecules – the largest found up to date is slightly higher than 100 atoms. The maximum such a speculation can say is that we have 'elective affinities' – preferences that are chemically determined – towards some location or some person. Energy – effort - must be used to break this bond, this preference. There are other interesting domains of research of course, one of them being genetics.

DNA is a mighty important molecule. Without it, there would be no such thing as a human (molecule). Those who try to create a human simply by synthesizing the calculated formula will see something unrelated to a human – unless of course they conquer the task of knowing every single bond that each atom has.
Another important 'molecule' is our brain. Throughout this book, two opposing views have been brought forward. On the one hand, the stance of Libb Thims was one that did not much take into account of neural computations. On the other hand, I have stressed that the brain is the central executive of our body – although this in no way implies free will. The true answer however lies in a combination of both: our brain is an environment itself in which physical happenings cause actions throughout our body, but our brain is determined by external forces as well – and perhaps mainly, since the former 'brain happenings' are determined by what's happening outside (sound, light, heat, nutrients and so on).

With regard to morality, it seems there are some differences in opinion. Thims supposes that the difference in Gibbs free energy should be a guideline for behavior. The first thing that comes in mind is that we all should have free energy meters with us, for else we wouldn't know what moral behavior was. A more general comment however is that there seems to me, at present, no reason why the Gibbs free energy dG should dictate whether a behavior is moral or not. Even suppose we had a device that could track it, this notion does not at all relate to what is considered moral behavior. Are we going to free violent rapists just because their behavior did not result from an abnormality in the Lewis inequality?

The questions that are asked throughout the ages have been discussed in this book. However, this book has been very inconclusive in many respects. So, where are we now? Instead of giving definite answers – which may appeal some readers, similar to biblical teachings – we have identified key questions, and revealed some factors that would be necessary to even start such an answer. The matter is complex, and is most often one of definition.

In this final page, I also want to address the comment that some ferociously advocate, that human thermodynamics is pseudoscience. While it may be true, note that most approaches to human understanding are worse – without all the ferocious arguments against them. For instance, many psychology researchers got away with fake results or serious abuse of data for a long time without anybody noticing – even if its conclusions were absolutely absurd (such as 'listening to this song makes you three years younger'). In contrast, to even suggest that physical principles may apply to humans often automatically is violently attacked by many.

This part of anthropocentrism is where I have issues with – that we are seen as 'free', exempt from laws, just because we have a brain. Another point however, with regard to anthropocentrism is that it is necessary to do 'moral science'. If there is no anthropocentric purpose to science, what good is it?

I may have discussed many concepts in a rather vague and hasty way. This reflects that I am only just exploring many of these concepts myself, and that explaining it in detail would require another book. Further books of mine may further elaborate on some of the implicit assumptions that I may have made without much apparent reason.

Ultimately, the contribution of this book is that I have raised some questions that are relevant to the topic of hmolscience, and beyond. I hope this book has offered important methodological considerations in investigating the human kind. I have usually not claimed to have some definite answer, but rather I have proposed which possible answers or philosophical directions are competing with each other. The reader may take this book as an inspiration for further inquiry, rather than an absolute answer. In other words, I have pointed out some viable directions, without having provided an entire map.

Index

Affinity coupling: 99

Aging: 53-62;84

AI, artificial intelligence: 10;53;64;66

Alien civilization: 52

Animate: 3;12;63-65; 70; 86; 94; 98

Anthropocentr-ism,-ize ,Anthropomorph-ism,-ize : 5; 38; 53-66; 74-76; 79; 82; 97; 111

Automaton theory: 33-34 ; 74

Bergman, Torbern: 101-102

Biology: 13-62
    Biologically coherent: 43
    Biologically degrading: 37
    Non-biological: 7; 13-32 ; 43 ; 49 ; 53 ; 67-68 ; 84-85; 109-110
    Semi-biological: 13,15,21,25

Bipedalism: 6

BossensNonFiction: 4 ; 45

Bottom-up: 27; 43 ; 55-58 ; 62 ; 66; 72 ; 88

Bond: 10 ; 93-111
    Bound state: 3; 34; 41; 108-109
    (Non-)Covalent bonds: 10 ; 108-110
    Photon-electron bindings: 10; 108-109

Brain: 5-6; 10; 16; 29; 38-40; 47; 55; 63-97; 110
    hydro-carbon brain: 63-64

Cell: 8-9; 12-62
    Cell membrane: 18-19
    Cell-as-a-molecule: 9; 36

Cellular automata: 7; 40; 67; 78; 89-90

Chemical determinism: 27; 41

CHNOPS: 29; 33-37

Coccoid cyanobacteria: 25-28

Cognition, cognitive:6; 39-40; 55; 72; 78; 82; 94; 109

Complexity: 33-34; 40-45; 63; 69; 92-93
    complexity theory: 7; 80-81

Computation: 82; 85; 89-90; 110

Conway, John : 7
(see also The Game of Life):

Consciousness: 22; 51; 61; 63-92

Cosmic balance: 31; 70

Descartes, Réné: 74-75; 79; 84
(see also Automaton theory; Dualism)

Definition: 7-62; 111
    Effective definition (see also Heart) 37-39
    Scalar definition: 37-39

Determinism:5;11; 27; 31-32 ; 41-46; 66-80; 87; 97
(see also Laplace's Demon)

Digital philosophy : 53; 64;67; 89-90
(see also Leibniz ; Wolfram ; Turing)

Dualism: 74-75

Eddington, Arthur: 94

Elective affinities: 3; 98-103; 110

Emergence: 45-48; 50

Enthalpy: 59; 101; 103

Entropy: 5; 11; 14; 27; 48; 54-60; 73
  Neg-entropy/entropy reversal: 57-61;
  entropy in marriage: 5; 70-73

Equilibrium: 12-13; 31-32;49 ; 59 ; 99

Evolution: 6-7; 12-14; 16-22; 77-88; 94
  origin of life: 16-22

Erythrocytes: 14-17; 27

Feynman, Richard : 109-110
(see also Quantum electrodynamics)

Genes/genome: 109-111
  DNA/RNA: 8-9; 13-32; 38-44; 109-111
  Genopsych : 4, 86-88, 91
  (see also DMR Sekhar)

Gestalt, Gestaltpsychology: 37-44; 47; 61; 68-69; 102

Gibbs, Willard Josiah / Gibbs free energy: 3;11;49;56; 59-60;71-72;91;95; 97-108

Gladyshev, Georgi : 4;38;49-55; 96
(see also Hierarchical thermodynamics)

Goethe, Johann : 3-4; 10; 43; 48; 51; 71; 91-92;97-102
(see also Elective affinities; Physics-based morality)

Gottmann, John: 5; 70; 73

Heart :32; 37; 55-61; 66; 70 ; 73; 93; 103
(see also Effective definition; Macro-certainties)

Holism : 44-48; 96
(see also Emergence; Gestalt; Reductionism)

Immortality: 68; 84

Instinct: 109-110

Intelligence: 10;22; 42; 52-53; 63-92

Jones, William: 44

Formally undecidable: 90

Free will: 11;16; 18;20; 24; 31;37; 41-42; 55;63-92;
  Retinal model of molecular choice: 63

Gödel incompleteness: 89

Human molecule: 3
  Dihumanide: 95
  Human molecular orbitals:3;6;9-10; 72 ;109-110

Hmolpedia: 4-7

Kaku, Michio: 76

Laplace's demon: 74; 77-80

Leibniz, Gottfried: 89

Lewis inequality: 86; 98

Life: 12-62
  Cosmic balance: 31; 70
  Half-life: 53-54
  Life force: 55; 75-79

Macro-certainties: 69-70

Marriage: 5;10; 53-54; 70-73; 97-106(see also Gottmann, John)

Mayr, Ernst : 45-51; 70
(see also Unbridgeable gap)

Metaphysics: 67; 83-90

Mills, Randell; 81

Mind-file: 39; 61

Molecule
    Human molecule: 3-9; 32; 39-44;51-55; 61
        Cell-as-a-molecule: 9
        Retinal molecule: 33; 63-64; 94

Morality: 97-107
    Moral monkeys: 104
    Negative confessions: 103
    Physics-based morality: 3-4; 97-107

Neg-entropy: 57-60

Neuroscience: 70-73

Non-covalent bonds: 109-111

Ockham, Edward: 93-94

Origin of life: 7; 12-62

Ostwald, Wilhelm:100

Pauling, Linus: 58-60

Phospholipids: 19

Plato's idea world: 83

Program:52;67;77; 85-90

Pythagorean philosophy: 90

Quantum physics:79-81
    Copenhagen interpretation: 80; 89
    Quantum electrodynamics: 109-110
    Uncertainty principle: 78
(see also Randomness)

Ra theology: 103

Randomness: 76-80

Reductionism: 44-48

Rutherford, Ernest: 53-54

Schrödinger, Erwin: 57-60
    Schrödinger's cat: 79

Sekhar, DMR : 4; 85-88
(see also Genopsych)

Sidis, William James: 54; 57-60

Singularity, singularitarian: 68; 84

Soul: see life debate, morality

Synthesis: 18; 43-44; 48-49; 53-56

Tautological: 104-106

Tesla, Nikola: 3; 20-21; 28-34; 42; 55;61; 69-70

The Game of Life: 7
(see also John Conway)

Thermodynamics
    Hierarchical thermodynamics: 4
    Human thermodynamics: 3
    Thermodynamic potential: 58-60

Thims, Libb: 3-4

Top-down: 56; 88

Tuhtan, Jeffrey: 4; 44-48

Turing
    Turing complete: 89
    Turing machine: 89

Unbridgeable gap: 34; 48-52
(see also Mayr, Ernst):

Uncertainty principle:78

Varicose veins: 32

Virus:
    erythrocytal-virus escape theory: 17

What is life?: 57-60

Wolfram, Stephen: 89-90

www.ingramcontent.com/pod-product-compliance
Lightning Source LLC
Chambersburg PA
CBHW072215170526
45158CB00002BA/617